Student Solutions Manual

Jay R. Schaffer

University of Northern Colorado

FOURTH EDITION

Elementary Statistics
Picturing *the* World

Larson | Farber

PEARSON

Prentice Hall

Upper Saddle River, NJ 07458

Editor-in-Chief, Mathematics & Statistics: Deirdre Lynch
Print Supplement Editor: Joanne Wendelken
Senior Managing Editor: Linda Mihatov Behrens
Project Manager, Production: Kristy S. Mosch
Art Director: Heather Scott
Supplement Cover Manager: Paul Gourhan
Supplement Cover Designer: Victoria Colotta
Operations Specialist: Ilene Kahn

© 2009 Pearson Education, Inc.
Pearson Prentice Hall
Pearson Education, Inc.
Upper Saddle River, NJ 07458

The author and publisher of this book have used their best efforts in preparing this book. These efforts include the development, research, and testing of the theories and programs to determine their effectiveness. The author and publisher make no warranty of any kind, expressed or implied, with regard to these programs or the documentation contained in this book. The author and publisher shall not be liable in any event for incidental or consequential damages in connection with, or arising out of, the furnishing, performance, or use of these programs.

Printed in the United States of America

10   9   8   7   6   5

ISBN-13:   978-0-13-601307-5

ISBN-10:   0-13-601307-4

Pearson Education Ltd., *London*
Pearson Education Australia Pty. Ltd., *Sydney*
Pearson Education Singapore, Pte. Ltd.
Pearson Education North Asia Ltd., *Hong Kong*
Pearson Education Canada, Inc., *Toronto*
Pearson Educación de Mexico, S.A. de C.V.
Pearson Education—Japan, *Tokyo*
Pearson Education Malaysia, Pte. Ltd.

CONTENTS

CONTENTS

Introduction to Statistics

1.1 AN OVERVIEW OF STATISTICS

1.1 Try It Yourself Solutions

1a. The population consists of the prices per gallon of regular gasoline at all gasoline stations in the United States.

b. The sample consists of the prices per gallon of regular gasoline at the 800 surveyed stations.

c. The data set consists of the 800 prices.

2a. Because the numerical measure of $2,326,706,685 is based on the entire collection of player's salaries, it is from a population.

b. Because the numerical measure is a characteristic of a population, it is a parameter.

3a. Descriptive statistics involve the statement "76% of women and 60% of men had a physical examination within the previous year."

b. An inference drawn from the study is that a higher percentage of women had a physical examination within the previous year.

1.1 EXERCISE SOLUTIONS

1. A sample is a subset of a population.

3. A parameter is a numerical description of a population characteristic. A statistic is a numerical description of a sample characteristic.

5. False. A statistic is a numerical measure that describes a sample characteristic.

7. True

9. False. A population is the collection of *all* outcomes, responses, measurements, or counts that are of interest.

11. The data set is a population because it is a collection of the ages of all the members of the House of Representatives.

13. The data set is a sample because the collection of the 500 spectators is a subset within the population of the stadium's 42,000 spectators.

15. Sample, because the collection of the 20 patients is a subset within the population.

17. Population: Party of registered voters in Warren County.

Sample: Party of Warren County voters responding to phone survey.

19. Population: Ages of adults in the United States who own computers.

Sample: Ages of adults in the United States who own Dell computers.

21. Population: All adults in the United States that take vacations.

Sample: Collection of 1000 adults surveyed that take vacations.

23. Population: Collection of all households in the U.S.

Sample: Collection of 1906 households surveyed.

25. Population: Collection of all registered voters.

Sample: Collection of 1045 registered voters surveyed.

27. Population: Collection of all women in the U.S.

Sample: Collection of the 546 U.S. women surveyed.

29. Statistic. The value $68,000 is a numerical description of a sample of annual salaries.

31. Parameter. The 62 surviving passengers out of 97 total passengers is a numerical description of all of the passengers of the Hindenburg that survived.

33. Statistic. 8% is a numerical description of a sample of computer users.

35. Statistic. 53% is a numerical description of a sample of people in the United States.

37. The statement "56% are the primary investor in their household" is an application of descriptive statistics.

An inference drawn from the sample is that an association exists between U.S. women and being the primary investor in their household.

39. Answers will vary.

41. (a) An inference drawn from the sample is that senior citizens who live in Florida have better memory than senior citizens who do not live in Florida.

(b) It implies that if you live in Florida, you will have better memory.

43. Answers will vary.

1.2 DATA CLASSIFICATION

1.2 Try It Yourself Solutions

1a. One data set contains names of cities and the other contains city populations.

b. City: Nonnumerical
Population: Numerical

c. City: Qualitative
Population: Quantitative

2a. (1) The final standings represent a ranking of basketball teams.

(2) The collection of phone numbers represents labels. No mathematical computations can be made.

b. (1) Ordinal, because the data can be put in order.

(2) Nominal, because you cannot make calculations on the data.

3a. (1) The data set is the collection of body temperatures.

(2) The data set is the collection of heart rates.

b. (1) Interval, because the data can be ordered and meaningful differences can be calculated, but it does not make sense writing a ratio using the temperatures.

  (2) Ratio, because the data can be ordered, can be written as a ratio, you can calculate meaningful differences, and the data set contains an inherent zero.

1.2 EXERCISE SOLUTIONS

1. Nominal and ordinal

3. False. Data at the ordinal level can be qualitative or quantitative.

5. False. More types of calculations can be performed with data at the interval level than with data at the nominal level.

7. Qualitative, because telephone numbers are merely labels.

9. Quantitative, because the lengths of songs on an MP3 player are numerical measures.

11. Qualitative, because the poll results are merely responses.

13. Qualitative. Ordinal. Data can be arranged in order, but differences between data entries make no sense.

15. Qualitative. Nominal. No mathematical computations can be made and data are categorized using names.

17. Qualitative. Ordinal. The data can be arranged in order, but differences between data entries are not meaningful.

19. Ordinal

21. Nominal

23. (a) Interval   (b) Nominal   (c) Ratio   (d) Ordinal

25. An inherent zero is a zero that implies "none." Answers will vary.

1.3 EXPERIMENTAL DESIGN

1.3 Try It Yourself Solutions

1a. (1) Focus: Effect of exercise on relieving depression.

  (2) Focus: Success of graduates.

  b. (1) Population: Collection of all people with depression.

  (2) Population: Collection of all university graduates.

  c. (1) Experiment

  (2) Survey

2a. There is no way to tell why people quit smoking. They could have quit smoking either from the gum or from watching the DVD.

  b. Two experiments could be done; one using the gum and the other using the DVD.

3a. Example: start with the first digits 92630782 . . .

b. 92|63|07|82|40|19|26

c. 63, 7, 40, 19, 26

4a. (1) The sample was selected by only using available students.

(2) The sample was selected by numbering each student in the school, randomly choosing a starting number, and selecting students at regular intervals from the starting number.

b. (1) Because the students were readily available in your class, this is convenience sampling.

(2) Because the students were ordered in a manner such that every 25th student is selected, this is systematic sampling.

1.3 EXERCISE SOLUTIONS

1. In an experiment, a treatment is applied to part of a population and responses are observed. In an observational study, a researcher measures characteristics of interest of part of a population but does not change existing conditions.

3. Assign numbers to each member of the population and use a random number table or use a random number generator.

5. True

7. False. Using stratified sampling guarantees that members of each group within a population will be sampled.

9. False. To select a systematic sample, a population is ordered in some way and then members of the population are selected at regular intervals.

11. In this study, you want to measure the effect of a treatment (using a fat substitute) on the human digestive system. So, you would want to perform an experiment.

13. Because it is impractical to create this situation, you would want to use a simulation.

15. (a) The experimental units are the 30–35 year old females being given the treatment.

(b) One treatment is used.

(c) A problem with the design is that there may be some bias on the part of the researchers if he or she knows which patients were given the real drug. A way to eliminate this problem would be to make the study into a double-blind experiment.

(d) The study would be a double-blind study if the researcher did not know which patients received the real drug or the placebo.

17. Each U.S. telephone number has an equal chance of being dialed and all samples of 1599 phone numbers have an equal chance of being selected, so this is a simple random sample. Telephone sampling only samples those individuals who have telephones, are available, and are willing to respond, so this is a possible source of bias.

19. Because the students were chosen due to their convenience of location (leaving the library), this is a convenience sample. Bias may enter into the sample because the students sampled may not be representative of the population of students. For example, there may be an association between time spent at the library and drinking habits.

21. Because a random sample of out-patients were selected and all samples of 1210 patients had an equal chance of being selected, this is a simple random sample.

23. Because a sample is taken from each one-acre subplot (stratum), this is a stratified sample.

25. Because every ninth name on a list is being selected, this is a systematic sample.

27. Answers will vary.

29. Census, because it is relatively easy to obtain the salaries of the 50 employees.

31. Question is biased because it already suggests that drinking fruit juice is good for you. The question might be rewritten as "How does drinking fruit juice affect your health?"

33. Question is unbiased because it does not imply how many hours of sleep are good or bad.

35. The households sampled represent various locations, ethnic groups, and income brackets. Each of these variables is considered a stratum.

37. Open Question
Advantage: Allows respondent to express some depth and shades of meaning in the answer.
Disadvantage: Not easily quantified and difficult to compare surveys.

    Closed Question
Advantage: Easy to analyze results.
Disadvantage: May not provide appropriate alternatives and may influence the opinion of the respondent.

39. Answers will vary.

41. The Hawthorne effect occurs when a subject changes behavior because he or she is in an experiment. However, the placebo effect occurs when a subject reacts favorably to a placebo he or she has been given.

43. Answers will vary.

CHAPTER 1 REVIEW EXERCISE SOLUTIONS

1. Population: Collection of all U.S. adults.

    Sample: Collection of the 1000 U.S. adults that were sampled.

3. Population: Collection of all credit cards.

    Sample: Collection of 146 credit cards that were sampled.

5. The team payroll is a parameter since it is a numerical description of a population (entire baseball team) characteristic.

7. Since "10 students" is describing a characteristic of a population of math majors, it is a parameter.

9. The average late fee of $27.46 charged by credit cards is representative of the descriptive branch of statistics. An inference drawn from the sample is that all credit cards charge a late fee of $27.46.

11. Quantitative because monthly salaries are numerical measurements.

13. Quantitative because ages are numerical measurements.

15. Interval. It makes no sense saying that 100 degrees is twice as hot as 50 degrees.

17. Nominal. The data are qualitative and cannot be arranged in a meaningful order.

19. Because CEOs keep accurate records of charitable donations, you could take a census.

21. In this study, you want to measure the effect of a treatment (fertilizer) on a soybean crop. You would want to perform an experiment.

23. The subjects could be split into male and female and then be randomly assigned to each of the five treatment groups.

25. Because random telephone numbers were generated and called, this is a simple random sample.

27. Because each community is considered a cluster and every pregnant woman in a selected community is surveyed, this is a cluster sample.

29. Because grade levels are considered strata and 25 students are sampled from each stratum, this is a stratified sample.

31. Telephone sampling only samples individuals who have telephones, are available, and are willing to respond.

33. The selected communities may not be representative of the entire area.

CHAPTER 1 QUIZ SOLUTIONS

1. Population: Collection of all individuals with anxiety disorders.

   Sample: Collection of 372 patients in study.

2. (a) Statistic. 19% is a characteristic of a sample of Internet users.

   (b) Parameter. 84% is a characteristic of the entire company (population).

   (c) Statistic. 40% is a characteristic of a sample of Americans.

3. (a) Qualitative, since post office box numbers are merely labels.

   (b) Quantitative, since a final exam is a numerical measure.

4. (a) Nominal. Badge numbers may be ordered numerically, but there is no meaning in this order and no mathematical computations can be made.

   (b) Ratio. It makes sense to say that the number of candles sold during the 1st quarter was twice as many as sold in the 2nd quarter.

   (c) Interval because meaningful differences between entries can be calculated, but a zero entry is not an inherent zero.

5. (a) In this study, you want to measure the effect of a treatment (low dietary intake of vitamin C and iron) on lead levels in adults. You want to perform an experiment.

   (b) Because it would be difficult to survey every individual within 500 miles of your home, sampling should be used.

6. Randomized Block Design

7. (a) Because people were chosen due to their convenience of location (on the campground), this is a convenience sample.

   (b) Because every tenth part is selected from an assembly line, this is a systematic sample.

   (c) Stratified sample because the population is first stratified and then a sample is collected from each stratum.

8. Convenience

Descriptive Statistics

2.1 Try It Yourself Solutions

1a. The number of classes (8) is stated in the problem.

b. Min = 15 Max = 89 Class width = $\dfrac{(89 - 15)}{8} = 9.25 \Rightarrow 10$

c.

Lower limit	Upper limit
15	24
25	34
35	44
45	54
55	64
65	74
75	84
85	94

d. See part (e).

e.

Class	Frequency, f
15–24	16
25–34	34
35–44	30
45–54	23
55–64	13
65–74	2
75–84	0
85–94	1

2a. See part (b).

b.

Class	Frequency, f	Midpoint	Relative frequency	Cumulative frequency
15–24	16	19.5	0.13	16
25–34	34	29.5	0.29	50
35–44	30	39.5	0.25	80
45–54	23	49.5	0.19	103
55–64	13	59.5	0.11	116
65–74	2	69.5	0.02	118
75–84	0	79.5	0.00	118
85–94	1	89.5	0.01	119
	$\sum f = 119$		$\sum \dfrac{f}{n} = 1$	

c. 86% of the teams scored fewer than 55 touchdowns. 3% of the teams scored more than 65 touchdowns.

3a.

Class Boundaries
14.5–24.5
24.5–34.5
34.5–44.5
44.5–54.5
54.5–64.5
64.5–74.5
74.5–84.5
84.5–94.5

b. Use class midpoints for the horizontal scale and frequency for the vertical scale.

c.

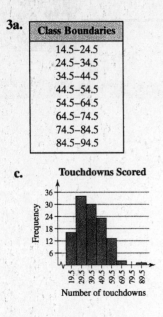

d. 86% of the teams scored fewer than 55 touchdowns. 3% of the teams scored more than 65 touchdowns.

4a. Use class midpoints for the horizontal scale and frequency for the vertical scale.

b. See part (c).

c.

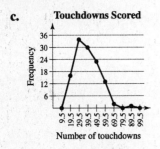

d. The number of touchdowns increases until 34.5 touchdowns, then decreases afterward.

5abc.

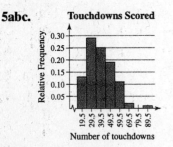

6a. Use upper class boundaries for the horizontal scale and cumulative frequency for the vertical scale.

b. See part (c).

c.

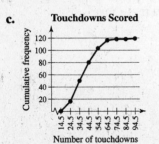

d. Approximately 80 teams scored 44 or fewer touchdowns.

e. Answers will vary.

7ab.

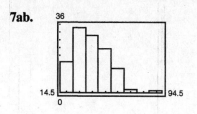

2.1 EXERCISE SOLUTIONS

1. By organizing the data into a frequency distribution, patterns within the data may become more evident.

3. Class limits determine which numbers can belong to that class.

Class boundaries are the numbers that separate classes without forming gaps between them.

5. False. Class width is the difference between the lower and upper limits of consecutive classes.

7. False. An ogive is a graph that displays cumulative frequency.

9. Width $= \dfrac{\text{Max} - \text{Min}}{\text{Classes}} = \dfrac{58 - 7}{6} = 8.5 \Rightarrow 9$

Lower class limits: 7, 16, 25, 34, 43, 52

Upper class limits: 15, 24, 33, 42, 51, 60

11. Width $= \dfrac{\text{Max} - \text{Min}}{\text{Classes}} = \dfrac{123 - 15}{6} = 18 \Rightarrow 19$

Lower class limits: 15, 34, 53, 72, 91, 110

Upper class limits: 33, 52, 71, 90, 109, 128

13. (a) Class width $= 31 - 20 = 11$

(b) and (c)

Class	Frequency, f	Midpoint	Class boundaries
20–30	19	25	19.5–30.5
31–41	43	36	30.5–41.5
42–52	68	47	41.5–52.5
53–63	69	58	52.5–63.5
64–74	74	69	63.5–74.5
75–85	68	80	74.5–85.5
86–96	24	91	85.5–96.5
	$\sum f = 365$		

15.

Class	Frequency, f	Midpoint	Relative frequency	Cumulative frequency
20–30	19	25	0.05	19
31–41	43	36	0.12	62
42–52	68	47	0.19	130
53–63	69	58	0.19	199
64–74	74	69	0.20	273
75–85	68	80	0.19	341
86–96	24	91	0.07	365
	$\sum f = 365$		$\sum \dfrac{f}{n} = 1$	

17. (a) Number of classes = 7

   (b) Least frequency $\approx$ 10

   (c) Greatest frequency $\approx$ 300

   (d) Class width = 10

19. (a) 50   (b) 22.5–24.5 lbs

21. (a) 24   (b) 29.5 lbs

23. (a) Class with greatest relative frequency: 8–9 inches.
   Class with least relative frequency: 17–18 inches.

   (b) Greatest relative frequency $\approx$ 0.195
   Least relative frequency $\approx$ 0.005

   (c) Approximately 0.015

25. Class with greatest frequency: 500–550
   Class with least frequency: 250–300 and 700–750

27. Class width $= \dfrac{\text{Max} - \text{Min}}{\text{Number of classes}} = \dfrac{39 - 0}{5} = 7.8 \Rightarrow 8$

Class	Frequency, f	Midpoint	Relative frequency	Cumulative frequency
0–7	8	3.5	0.32	8
8–15	8	11.5	0.32	16
16–23	3	19.5	0.12	19
24–31	3	27.5	0.12	22
32–39	3	35.5	0.12	25
	$\sum f = 25$		$\sum \dfrac{f}{n} = 1$	

Class with greatest frequency: 0–7, 8–15
Class with least frequency: 16–23, 24–31, 32–39

29. Class width $= \dfrac{\text{Max} - \text{Min}}{\text{Number of classes}} = \dfrac{7119 - 1000}{6} = 1019.83 \Rightarrow 1020$

Class	Frequency, f	Midpoint	Relative frequency	Cumulative frequency
1000–2019	12	1509.5	0.5455	12
2020–3039	3	2529.5	0.1364	15
3040–4059	2	3549.5	0.0909	17
4060–5079	3	4569.5	0.1364	20
5080–6099	1	5589.5	0.0455	21
6100–7119	1	6609.5	0.0455	22
	$\sum f = 22$		$\sum \dfrac{f}{n} = 1$	

July Sales for Representatives

Class with greatest frequency: 1000–2019
Class with least frequency: 5080–6099; 6100–7119

31. Class width $= \dfrac{\text{Max} - \text{Min}}{\text{Number of classes}} = \dfrac{514 - 291}{8} = 27.875 \Rightarrow 28$

Class	Frequency, f	Midpoint	Relative frequency	Cumulative frequency
291–318	5	304.5	0.1667	5
319–346	4	332.5	0.1333	9
347–374	3	360.5	0.1000	12
375–402	5	388.5	0.1667	17
403–430	6	416.5	0.2000	23
431–458	4	444.5	0.1333	27
459–486	1	472.5	0.0333	28
487–514	2	500.5	0.0667	30
	$\sum f = 30$		$\sum \dfrac{f}{n} = 1$	

Class with greatest frequency: 403–430

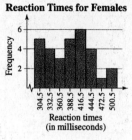

Reaction Times for Females

Reaction times (in milliseconds)

33. Class width $= \dfrac{\text{Max} - \text{Min}}{\text{Number of classes}} = \dfrac{264 - 146}{5} = 23.6 \Rightarrow 24$

Class	Frequency, f	Midpoint	Relative frequency	Cumulative frequency
146–169	6	157.5	0.2308	6
170–193	9	181.5	0.3462	15
194–217	3	205.5	0.1154	18
218–241	6	229.5	0.2308	24
242–265	2	253.5	0.0769	26
	$\sum f = 26$		$\sum \dfrac{f}{n} = 1$	

Class with greatest relative frequency: 170–193
Class with least relative frequency: 242–265

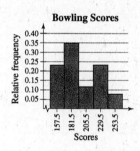

Bowling Scores

Scores

35. Class width $= \dfrac{\text{Max} - \text{Min}}{\text{Number of classes}} = \dfrac{52 - 33}{5} = 3.8 \Rightarrow 4$

Class	Frequency, f	Midpoint	Relative frequency	Cumulative frequency
33–36	8	34.5	0.3077	8
37–40	6	38.5	0.2308	14
41–44	5	42.5	0.1923	19
45–48	2	46.5	0.0769	21
49–52	5	50.5	0.1923	26
	$\sum f = 26$		$\sum \dfrac{f}{n} = 1$	

Class with greatest relative frequency: 33–36
Class with least relative frequency: 45–48

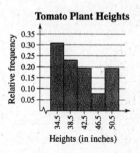

Tomato Plant Heights

Heights (in inches)

37. Class width $= \dfrac{\text{Max} - \text{Min}}{\text{Number of classes}} = \dfrac{73 - 52}{6} = 3.5 \Rightarrow 4$

Class	Frequency, f	Relative frequency	Cumulative frequency
52–55	3	0.125	3
56–59	3	0.125	6
60–63	9	0.375	15
64–67	4	0.167	19
68–71	4	0.167	23
72–75	1	0.042	24
	$\sum f = 24$	$\sum \dfrac{f}{n} \approx 1$	

Location of the greatest increase in frequency: 60–63

Retirement Ages

39. Class width $= \dfrac{\text{Max} - \text{Min}}{\text{Number of classes}} = \dfrac{18 - 2}{6} = 2.67 \Rightarrow 3$

Class	Frequency, f	Relative frequency	Cumulative frequency
2–4	9	0.3214	9
5–7	6	0.2143	15
8–10	7	0.2500	22
11–13	3	0.1071	25
14–16	2	0.0714	27
17–19	1	0.0357	28
	$\sum f = 28$	$\sum \dfrac{f}{n} \approx 1$	

Location of the greatest increase in frequency: 2–4

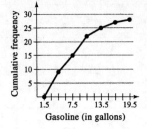

Gallons of Gasoline Purchased

41. Class width $= \dfrac{\text{Max} - \text{Min}}{\text{Number of classes}} = \dfrac{98 - 47}{5} = 10.2 \Rightarrow 11$

Class	Frequency, f	Midpoint	Relative frequency	Cumulative frequency
47–57	1	52	0.05	1
58–68	1	63	0.05	2
69–79	5	74	0.25	7
80–90	8	85	0.40	15
91–101	5	96	0.25	20
	$\sum f = 20$		$\sum \dfrac{f}{N} = 1$	

Class with greatest frequency: 80–90
Classes with least frequency: 47–57 and 58–68

Exam Scores

43. (a) Class width $= \dfrac{\text{Max} - \text{Min}}{\text{Number of classes}} = \dfrac{104 - 61}{8} = 5.375 \Rightarrow 6$

Class	Frequency, f	Midpoint	Relative frequency
61–66	1	63.5	0.0333
67–72	3	69.5	0.1000
73–78	6	75.5	0.2000
79–84	10	81.5	0.3333
85–90	5	87.5	0.1667
91–96	2	93.5	0.0667
97–102	2	99.5	0.0667
103–108	1	105.5	0.0333
	$\sum f = 30$		$\sum \dfrac{f}{N} = 1$

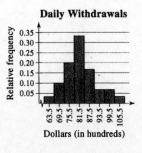

Daily Withdrawals

(b) 16.7%, because the sum of the relative frequencies for the last three classes is 0.167.

(c) $9600, because the sum of the relative frequencies for the last two classes is 0.10.

45.

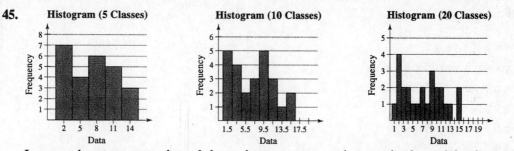

In general, a greater number of classes better preserves the actual values of the data set, but is not as helpful for observing general trends and making conclusions. When choosing the number of classes, an important consideration is the size of the data set. For instance, you would not want to use 20 classes if your data set contained 20 entries. In this particular example, as the number of classes increases, the histogram shows more fluctuation. The histograms with 10 and 20 classes have classes with zero frequencies. Not much is gained by using more than five classes. Therefore, it appears that five classes would be best.

2.2 MORE GRAPHS AND DISPLAYS

2.2 Try It Yourself Solutions

1a. 1
  2
  3
  4
  5
  6
  7
  8

b. Key: 1|7 = 17

  1 | 7 5 8 8 5 5
  2 | 7 6 8 9 8 9 7 9 8 7 5 3 4 6 2 5 0 1 1 2 1 4 1
  3 | 9 9 7 7 8 8 6 4 7 6 5 5 5 1 4 5 9 8 2 5 2 2 2 3 3 3 2 4 1 1 0 4 2 1 2 1 0
  4 | 9 8 6 8 7 8 6 4 8 5 4 6 6 7 1 1 5 4 5 3 2 2 8 3 0 4 0 5 3 0
  5 | 9 4 5 4 3 5 5 9 0 2 3 5 7 0 5
  6 | 8 5 1 3 3 1 1 0
  7 |
  8 | 9

c. Key: 1|7 = 17

  1 | 5 5 5 7 8 8
  2 | 0 1 1 1 1 2 2 3 4 4 5 5 6 6 7 7 7 8 8 8 9 9 9
  3 | 0 0 1 1 1 1 1 2 2 2 2 2 2 3 3 3 4 4 4 4 5 5 5 5 5 6 6 7 7 7 8 8 8 9 9 9
  4 | 0 0 0 1 1 2 2 3 3 3 3 4 4 4 4 5 5 5 5 6 6 6 6 7 7 8 8 8 8 8 9
  5 | 0 0 2 3 3 4 4 5 5 5 5 7 9 9
  6 | 0 1 1 1 3 3 5 8
  7 |
  8 | 9

d. It seems that most teams scored under 54 touchdowns.

2ab. Key: $1|7 = 17$

```
1 |
1 | 5 5 5 7 8 8
2 | 0 1 1 1 1 2 2 3 4 4
2 | 5 5 6 6 7 7 7 8 8 8 9 9 9
3 | 0 0 1 1 1 1 1 2 2 2 2 2 2 2 3 3 3 4 4 4 4
3 | 5 5 5 5 5 6 6 7 7 7 8 8 8 9 9 9
4 | 0 0 0 1 1 2 2 3 3 3 4 4 4 4 4
4 | 5 5 5 5 6 6 6 6 7 7 8 8 8 8 8 9
5 | 0 0 2 3 3 4 4
5 | 5 5 5 5 7 9 9
6 | 0 1 1 1 3 3
6 | 5 8
7 |
7 |
8 |
8 | 9
```

3a. Use number of touchdowns for the horizontal axis.

b. **Touchdowns Scored**

Number of Touchdowns

c. It appears that a large percentage of teams scored under 50 touchdowns.

4a.

Vehicle type	Killed (frequency)	Relative frequency	Central angle
Cars	22,423	0.64	$(0.64)(360°) \approx 230°$
Trucks	10,216	0.29	$(0.29)(360°) \approx 104°$
Motorcycles	2,227	0.06	$(0.06)(360°) \approx 22°$
Other	425	0.01	$(0.01)(360°) \approx 4°$
	$\sum f = 35{,}291$	$\sum \frac{f}{n} \approx 1$	$\sum = 360°$

b. **Motor Vehicle Occupants Killed in 1995**

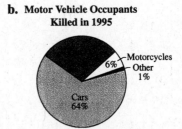

Motorcycles 6%
Other 1%
Cars 64%

c. As a percentage of total vehicle deaths, car deaths decreased by 15%, truck deaths increased by 8%, and motorcycle deaths increased by 6%.

5a.

Cause	Frequency, f
Auto Dealers	14,668
Auto Repair	9,728
Home Furnishing	7,792
Computer Sales	5,733
Dry Cleaning	4,649

b. **Causes of BBB Complaints**

c. It appears that the auto industry (dealers and repair shops) account for the largest portion of complaints filed at the BBB.

6ab.

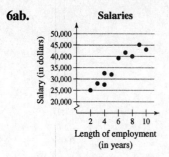

Salaries

c. It appears that the longer an employee is with the company, the larger his/her salary will be.

7ab.

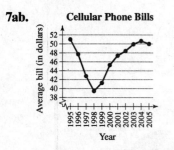

Cellular Phone Bills

c. It appears that the average monthly bill for cellular telephone subscribers decreased significantly from 1995 to 1998, then increased from 1998 to 2004.

2.2 EXERCISE SOLUTIONS

1. Quantitative: Stem-and-Leaf Plot, Dot Plot, Histogram, Time Series Chart, Scatter Plot

Qualitative: Pie Chart, Pareto Chart

3. Both the stem-and-leaf plot and the dot plot allow you to see how data are distributed, determine specific data entries, and identify unusual data values.

5. b   **7.** a

9. 27, 32, 41, 43, 43, 44, 47, 47, 48, 50, 51, 51, 52, 53, 53, 53, 54, 54, 54, 54, 55, 56, 56, 58, 59, 68, 68, 68, 73, 78, 78, 85

Max: 85   Min: 27

11. 13, 13, 14, 14, 14, 15, 15, 15, 15, 15, 16, 17, 17, 18, 19

Max: 19   Min: 13

13. Anheuser-Busch is the top sports advertiser spending approximately $190 million. Honda spends the least. (Answers will vary.)

15. Tailgaters irk drivers the most, while too cautious drivers irk drivers the least. (Answers will vary.)

17. Key: 6|7 = 67

```
6 | 7 8
7 | 3 5 5 6 9
8 | 0 0 2 3 5 5 7 7 8
9 | 0 1 1 1 2 4 5 5
```

Most grades of the biology midterm were in the 80s or 90s.

19. Key: 4|3 = 4.3

```
4 | 3 9
5 | 1 8 8 8 9
6 | 4 8 9 9 9
7 | 0 0 2 2 2 5
8 | 0 1
```

It appears that most ice had a thickness of 5.8 centimeters to 7.2 centimeters.
(Answers will vary.)

21.

Advertisements

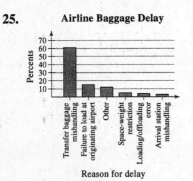

It appears that most of the 30 people from the U.S. see or hear between 450 and
750 advertisements per week. (Answers will vary.)

23.

Category	Frequency	Relative frequency	Angle
North America	23	0.12	(0.12)(360°) ≈ 43°
South America	12	0.06	(0.06)(360°) ≈ 23°
Europe	43	0.22	(0.22)(360°) ≈ 81°
Oceania	14	0.07	(0.07)(360°) ≈ 26°
Africa	53	0.28	(0.28)(360°) ≈ 99°
Asia	47	0.25	(0.25)(360°) ≈ 88°
	$\sum f = 192$	$\sum \frac{f}{n} = 1$	

Countries in the United Nations

Asia 25%
North America 12%
Europe 22%
Oceania 7%
South America 6%
Africa 28%

Most countries in the United Nations come from Africa and the least amount come from
South America. (Answers will vary.)

25.

Airline Baggage Delay

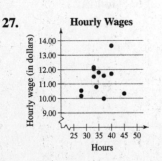

It appears that the biggest reason for baggage delay comes
from transfer baggage mishandling. (Answers will vary.)

27.

Hourly Wages

It appears that hourly wage increases as the number of hours
worked increases. (Answers will vary.)

29.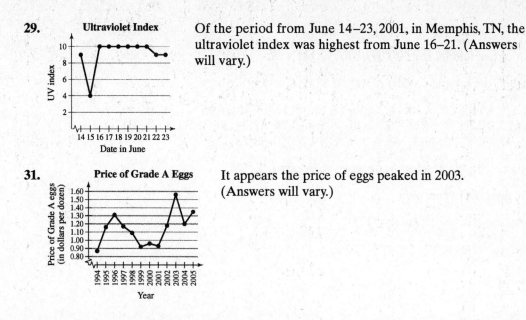

Of the period from June 14–23, 2001, in Memphis, TN, the ultraviolet index was highest from June 16–21. (Answers will vary.)

31.

It appears the price of eggs peaked in 2003. (Answers will vary.)

33. (a) When data are taken at regular intervals over a period of time, a time series chart should be used. (Answers will vary.)

(b)

35. (a) At law firm A, the lowest salary was $90,000 and the highest was $203,000; at law firm B, the lowest salary was $90,000 and the highest salary was $190,000.

(b) There are 30 lawyers at law firm A and 32 lawyers at law firm B.

(c) At Law Firm A, the salaries tend to be clustered at the far ends of the distribution range and at Law Firm B, the salaries tend to fall in the middle of the distribution range.

2.3 MEASURES OF CENTRAL TENDENCY

2.3 Try It Yourself Solutions

1a. $\Sigma x = 578$

b. $\bar{x} = \dfrac{\Sigma x}{n} = \dfrac{578}{14} = 41.3$

c. The mean age of an employee in a department is 41.3 years.

2a. 18 18, 19, 19, 19, 20, 21, 21, 21, 21, 23, 24, 24, 26, 27, 27, 29, 30, 30, 30, 33, 33, 34, 35, 38

b. median = middle entry = 24

c. The median age for the sample of fans at the concert is 24.

3a. 70, 80, 100, 130, 140, 150, 160, 200, 250, 270

b. median = mean of two middle entries {140, 150} = 145

c. The median price of the sample of MP3 players is $145.

4a. 0, 0, 1, 1, 1, 2, 3, 3, 3, 4, 5, 5, 5, 7, 9, 10, 12, 12, 13, 13, 13, 13, 13, 15, 16, 16, 17, 17, 18, 18, 18, 19, 19, 19, 20, 20, 21, 22, 23, 23, 24, 24, 25, 25, 26, 26, 26, 29, 33, 36, 37, 39, 39, 39, 39, 40, 40, 41, 41, 41, 42, 44, 44, 45, 47, 48, 49, 49, 49, 51, 53, 56, 58, 58, 59, 60, 67, 68, 68, 72

b. The age that occurs with the greatest frequency is 13 years old.

c. The mode of the ages is 13 years old.

5a. "Yes" occurs with the greatest frequency (171).

b. The mode of the responses to the survey is "Yes".

6a. $\bar{x} = \dfrac{\Sigma x}{n} = \dfrac{410}{19} \approx 21.6$

median = 21

mode = 20

b. The mean in Example 6 ($\bar{x} \approx 23.8$) was heavily influenced by the age 65. Neither the median nor the mode was affected as much by the age 65.

7ab.

Source	Score, x	Weight, w	$x \cdot w$
Test Mean	86	0.50	(83)(0.50) = 43.0
Midterm	96	0.15	(96)(0.15) = 14.4
Final	98	0.20	(98)(0.20) = 19.6
Computer Lab	98	0.10	(98)(0.10) = 9.8
Homework	100	0.05	(100)(0.05) = 5.0
		$\Sigma w = 1.00$	$\Sigma (x \cdot w) = 91.8$

c. $\bar{x} = \dfrac{\Sigma (x \cdot w)}{\Sigma w} = \dfrac{91.8}{1.00} = 91.8$

d. The weighted mean for the course is 91.8. So, you did get an A.

8abc.

Class	Midpoint, x	Frequency, f	$x \cdot f$
15–24	19.5	16	312
25–34	29.5	34	1003
35–44	39.5	30	1185
45–54	49.5	23	1138.5
55–64	59.5	13	773.5
65–74	69.5	2	139
75–84	79.5	0	0
85–94	89.5	1	89.5
		$N = 119$	$\Sigma (x \cdot f) = 4640.5$

d. $\mu = \dfrac{\Sigma (x \cdot f)}{N} = \dfrac{4640.5}{119} \approx 39.0$

The average number of touchdowns is approximately 39.0.

2.3 EXERCISE SOLUTIONS

1. True.

3. False. All quantitative data sets have a median.

5. False. When each data class has the same frequency, the distribution is uniform.

7. Answers will vary. A data set with an outlier within it would be an example. For instance, the mean of the prices of existing home sales tends to be "inflated" due to the presence of a few very expensive homes.

9. Skewed right because the "tail" of the distribution extends to the right.

11. Uniform because the bars are approximately the same height.

13. (11), because the distribution values range from 1 to 12 and has (approximately) equal frequencies.

15. (12), because the distribution has a maximum value of 90 and is skewed left due to a few students scoring much lower than the majority of the students.

17. $\bar{x} = \dfrac{\Sigma x}{n} = \dfrac{81}{13} \approx 6.2$

5 5 5 5 5 5 ⑥ 6 7 8 9 9
 └── middle value $\Rightarrow$ median = 6

mode = 5   (occurs 6 times)

19. $\bar{x} = \dfrac{\Sigma x}{n} = \dfrac{32}{7} \approx 4.57$

3.7 4.0 4.8 ④.8 4.8 4.8 5.1
 └── middle value $\Rightarrow$ median = 4.8

mode = 4.8   (occurs 4 times)

21. $\bar{x} = \dfrac{\Sigma x}{n} = \dfrac{661.2}{32} \approx 20.66$

10.5, 13.2, 14.9, 16.2, 16.7, 16.9, 17.6, 18.2, 18.6, 18.8, 18.8, 19.1, 19.2, 19.6, 19.8,

19.9, 20.2, 20.7, 20.9, 22.1, 22.1, 22.2, 22.9, 23.2, 23.3, 24.1, 24.9, 25.8, 26.6, 26.7, 26.7, 30.8
 └── two middle values $\Rightarrow$ median = $\dfrac{19.9 + 20.2}{2}$ = 20.05

mode = 18.8, 22.1, 26.7   (occurs 2 times each)

23. $\bar{x}$ = not possible (nominal data)

median = not possible (nominal data)

mode = "Worse"

The mean and median cannot be found because the data are at the nominal level of measurement.

25. $\bar{x} = \dfrac{\Sigma x}{n} = \dfrac{1194.4}{7} \approx 170.63$

155.7, 158.1, 162.2, ⟨169.3⟩, 180, 181.8, 187.3

⎣———— middle value $\Rightarrow$ median = 169.3

mode = none

The mode cannot be found because no data points are repeated.

27. $\bar{x} = \dfrac{\Sigma x}{n} = \dfrac{226}{10} = 22.6$

14, 14, 15, 177, 18, 20, 22, 25, 40, 41

⎣———— two middle values $\Rightarrow$ median = $\dfrac{18 + 20}{2}$ = 19

mode = 14   (occurs 2 times)

29. $\bar{x} = \dfrac{\Sigma x}{n} = \dfrac{197.5}{14} \approx 14.11$

1.5, 2.5, 2.5, 5, 10.5, 11, 13, 15.5, 16.5, 17.5, 20, 26.5, 27, 28.5

⎣———— two middle values $\Rightarrow$ median = $\dfrac{13 + 15.5}{2}$ = 14.25

mode = 2.5   (occurs 2 times)

31. $\bar{x} = \dfrac{\Sigma x}{n} = \dfrac{578}{14} = 41.3$

10, 12, 21, 24, 27, 37, 38, 41, 45, 45, 50, 57, 65, 106

⎣———— two middle values $\Rightarrow$ median = $\dfrac{38 + 41}{2}$ = 39.5

mode = 4.5   (occurs 2 times)

33. $\bar{x} = \dfrac{\Sigma x}{n} = \dfrac{292}{15} \approx 19.5$

5, 8, 10, 15, 15, 15, 17, ⟨20⟩ 21, 22, 22, 25, 28, 32, 37

⎣———— middle value $\Rightarrow$ median = 20

mode = 15   (occurs 3 times)

35. A = mode (data entry that occurred most often)

B = median (left of mean in skewed-right dist.)

C = mean (right of median in skewed-right dist.)

37. Mode because the data is nominal.

39. Mean because the data does not contain outliers.

41.

Source	Score, x	Weight, w	$x \cdot w$
Homework	85	0.05	$(85)(0.05) = 4.25$
Quiz	80	0.35	$(80)(0.35) = 28$
Project	100	0.20	$(100)(0.20) = 20$
Speech	90	0.15	$(90)(0.15) = 13.5$
Final Exam	93	0.25	$(93)(0.25) = 23.25$
		$\Sigma w = 1$	$\Sigma(x \cdot w) = 89$

$$\bar{x} = \frac{\Sigma(x \cdot w)}{\Sigma w} = \frac{89}{1} = 89$$

43.

Balance, x	Days, w	$x \cdot w$
$523	24	$(523)(24) = 12,552$
$2415	2	$(2415)(2) = 4830$
$250	4	$(250)(4) = 1000$
	$\Sigma w = 30$	$\Sigma(x \cdot w) = 18,382$

$$\bar{x} = \frac{\Sigma(x \cdot w)}{\Sigma w} = \frac{18,382}{30} = \$612.73$$

45.

Grade	Points, x	Credits, w	$x \cdot w$
B	3	3	$(3)(3) = 9$
B	3	3	$(3)(3) = 9$
A	4	4	$(4)(4) = 16$
D	1	2	$(1)(2) = 2$
C	2	3	$(2)(3) = 6$
		$\Sigma w = 15$	$\Sigma(x \cdot w) = 42$

$$\bar{x} = \frac{\Sigma(x \cdot w)}{\Sigma w} = \frac{42}{15} = 2.8$$

47.

Midpoint, x	Frequency, f	$x \cdot f$
61	4	$(61)(4) = 244$
64	5	$(64)(5) = 320$
67	8	$(67)(8) = 536$
70	1	$(70)(1) = 70$
	$n = 18$	$\Sigma(x \cdot f) = 1170$

$$\bar{x} = \frac{\Sigma(x \cdot f)}{n} = \frac{1170}{18} \approx 65 \text{ inches}$$

49.

Midpoint, x	Frequency, f	$x \cdot f$
4.5	55	$(4.5)(55) = 247.5$
14.5	70	$(14.5)(70) = 1015$
24.5	35	$(24.5)(35) = 857.5$
34.5	56	$(34.5)(56) = 1932$
44.5	74	$(44.5)(74) = 3293$
54.5	42	$(54.5)(42) = 2289$
64.5	38	$(64.5)(38) = 2451$
74.5	17	$(74.5)(17) = 1266.5$
84.5	10	$(84.5)(10) = 845$
	$n = 397$	$\Sigma(x \cdot f) = 14,196.5$

$$\bar{x} = \frac{\Sigma(x \cdot f)}{n} = \frac{14,196.5}{397} \approx 35.8 \text{ years old}$$

51. Class width $= \dfrac{\text{Max} - \text{Min}}{\text{Number of classes}} = \dfrac{14 - 3}{6} = 1.83 \Rightarrow 2$

Class	Midpoint, x	Frequency, f
3–4	3.5	3
5–6	5.5	8
7–8	7.5	4
9–10	9.5	2
11–12	11.5	2
13–14	13.5	1
		$\Sigma f = 20$

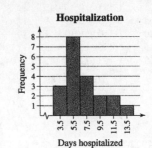

Shape: Positively skewed

53. Class width $= \dfrac{\text{Max} - \text{Min}}{\text{Number of classes}} = \dfrac{76 - 62}{5} = 2.8 \Rightarrow 3$

Class	Midpoint, x	Frequency, f
62–64	63	3
65–67	66	7
68–70	69	9
71–73	72	8
74–76	75	3
		$\Sigma f = 30$

Shape: Symmetric

55. (a) $\bar{x} = \dfrac{\Sigma x}{n} = \dfrac{36.03}{6} = 6.005$

5.59, 5.99, $\underbrace{6, \; 6.02}_{\text{two middle values}}$, 6.03, 6.4 $\Rightarrow$ median $= \dfrac{6 + 6.02}{2} = 6.01$

(b) $\bar{x} = \dfrac{\Sigma x}{n} = \dfrac{35.67}{6} = 5.945$

5.59, 5.99, $\underbrace{6, \; 6.02}_{\text{two middle values}}$, 6.03, 6.4 $\Rightarrow$ median $= \dfrac{6 + 6.02}{2} = 6.01$

(c) mean

57. (a) $\bar{x} = \dfrac{\Sigma x}{n} = \dfrac{3222}{9} = 358$

147, 177, 336, 360, $\overset{\text{middle value}}{\textcircled{375}}$ 393, 408, 504, 522 $\Rightarrow$ median $= 375$

(b) $\bar{x} = \dfrac{\Sigma x}{n} = \dfrac{9666}{9} = 1074$

441, 531, 1008, 1080, $\overset{\text{middle value}}{\textcircled{1125}}$ 1179, 1224, 1512, 1566 $\Rightarrow$ median $= 1125$

(c) $\bar{x}$ from part (b) is 3 times $\bar{x}$ from part (a). Median from part (b) is 3 times median from part (a).

(d) Multiply part (b) answers by 12.

59. Car A because it has the highest midrange of the three.

Car A: Midrange $= \dfrac{34 + 28}{2} = 32$

Car B: Midrange $= \dfrac{31 + 29}{2} = 30$

Car C: Midrange $= \dfrac{32 + 28}{2} = 30$

61. (a) Order the data values.

11 13 22 28 36 36 36 37 37 37 38 41 43 44 46
47 51 51 51 53 61 62 63 64 72 72 74 76 85 90

Delete the lowest 10%, smallest 3 observations (11, 13, 22).

Delete the highest 10%, largest 3 observations (76, 85, 90).

Find the 10% trimmed mean using the remaining 24 observations.

10% trimmed mean = 49.2

(b) $\bar{x} = 49.2$

median = 46.5

mode = 36, 37, 51

midrange = 50.5

(c) Using a trimmed mean eliminates potential outliers that may affect the mean of all the observations.

2.4 MEASURES OF VARIATION

2.4 Try It Yourself Solutions

1a. Min = 23 or \$23,000 and Max = 58 or \$58,000

b. Range = max − min = 58 − 23 = 35 or \$35,000

c. The range of the starting salaries for Corporation B is 35 or \$35,000 (much larger than range of Corporation A).

2a. $\mu = \dfrac{\Sigma x}{N} = \dfrac{415}{10} = 41.5$ or $\$41,500$

b.

Salary, x (1000s of dollars)	Deviation, $x - \mu$ (1000s of dollars)
23	$23 - 41.5 = -18.5$
29	$29 - 41.5 = -12.5$
32	$32 - 41.5 = -9.5$
40	$40 - 41.5 = -1.5$
41	$41 - 41.5 = -0.5$
41	$41 - 41.5 = -0.5$
49	$49 - 41.5 = 7.5$
50	$50 - 41.5 = 8.5$
52	$52 - 41.5 = 10.5$
58	$58 - 41.5 = 16.5$
$\sum x = 415$	$\sum(x - \mu) = 0$

3ab. $\mu = 41.5$ or $\$41,500$

Salary, x	$x - \mu$	$(x - \mu)^2$
23	-18.5	$(-18.5)^2 = 342.25$
29	-12.5	$(-12.5)^2 = 156.25$
32	-9.5	$(-9.5)^2 = 90.25$
40	-1.5	$(-1.5)^2 = 2.25$
41	-0.5	$(-0.5)^2 = 0.25$
41	-0.5	$(-0.5)^2 = 0.25$
49	7.5	$(7.5)^2 = 56.25$
50	8.5	$(8.5)^2 = 72.25$
52	10.5	$(10.5)^2 = 110.25$
58	16.5	$(16.5)^2 = 272.25$
$\sum x = 415$	$\sum(x - \mu) = 0$	$\sum(x - \mu)^2 = 1102.5$

c. $\sigma^2 = \dfrac{\Sigma(x - \mu)^2}{N} = \dfrac{1102.5}{10} \approx 110.25$

d. $\sigma = \sqrt{\sigma^2} = \sqrt{\dfrac{1102.5}{10}} = 10.5$ or $\$10,500$

e. The population variance is 110.3 and the population standard deviation is 10.5 or $\$10,500$.

4a.

Salary, x	$x - \bar{x}$	$(x - \bar{x})^2$
23	-18.5	342.25
29	-12.5	156.25
32	-9.5	90.25
40	-1.5	2.25
41	-0.5	0.25
41	-0.5	0.25
49	7.5	56.25
50	8.5	72.25
52	10.5	110.25
58	16.5	272.25
$\sum x = 415$	$\sum(x - \bar{x}) = 0$	$\sum(x - \bar{x})^2 = 1102.5$

$SS_x = \Sigma(x - \bar{x})^2 = 1102.5$

b. $s^2 = \dfrac{\Sigma(x - \bar{x})^2}{(n - 1)} = \dfrac{1102.5}{9} = 122.5$

c. $s = \sqrt{s^2} = \sqrt{122.5} \approx 11.1$ or $\$11,100$

5a. (Enter data in computer or calculator)

b. $\bar{x} = 37.89$, $s = 3.98$

6a. 7, 7, 7, 7, 7, 13, 13, 13, 13, 13

b.

x	$x - \mu$	$(x - \mu)^2$
7	$7 - 10 = -3$	$(-3)^2 = 9$
7	$7 - 10 = -3$	$(-3)^2 = 9$
7	$7 - 10 = -3$	$(-3)^2 = 9$
7	$7 - 10 = -3$	$(-3)^2 = 9$
7	$7 - 10 = -3$	$(-3)^2 = 9$
13	$13 - 10 = 3$	$(3)^2 = 9$
13	$13 - 10 = 3$	$(3)^2 = 9$
13	$13 - 10 = 3$	$(3)^2 = 9$
13	$13 - 10 = 3$	$(3)^2 = 9$
13	$13 - 10 = 3$	$(3)^2 = 9$
$\Sigma x = 100$	$\sum(x - \mu) = 0$	$\sum(x - \mu)^2 = 90$

$$\mu = \frac{\Sigma x}{N} = \frac{100}{10} = 10$$

$$\sigma = \sqrt{\frac{\Sigma(x - \mu)^2}{N}} = \sqrt{\frac{90}{10}} = \sqrt{9} = 3$$

7a. $64 - 61.25 = 2.75 = 1$ standard deviation

b. 34%

c. The estimated percent of the heights that are between 61.25 and 64 inches is 34%.

8a. $31.6 - 2(19.5) = -7.4 = \ > 0$

b. $31.6 + 2(19.5) = 70.6$

c. $1 - \frac{1}{k^2} = 1 - \frac{1}{(2)^2} = 1 - \frac{1}{4} = 0.75$

At least 75% of the data lie within 2 standard deviations of the mean. At least 75% of the population of Alaska is between 0 and 70.6 years old.

9a.

x	f	xf
0	10	$(0)(10) = 0$
1	19	$(1)(19) = 19$
2	7	$(2)(7) = 14$
3	7	$(3)(7) = 21$
4	5	$(4)(5) = 20$
5	1	$(5)(1) = 5$
6	1	$(6)(1) = 6$
	$n = 50$	$\sum xf = 85$

b. $\bar{x} = \frac{\Sigma xf}{n} = \frac{85}{50} = 1.7$

c.

$x - \bar{x}$	$(x - \bar{x})^2$	$(x - \bar{x})^2 \cdot f$
$0 - 1.7 = -1.70$	$(-1.70)^2 = 2.8900$	$(2.8900)(10) = 28.90$
$1 - 1.7 = -0.70$	$(-0.70)^2 = 0.4900$	$(0.4900)(19) = 9.31$
$2 - 1.7 = 0.30$	$(0.30)^2 = 0.0900$	$(0.0900)(7) = 0.63$
$3 - 1.7 = 1.30$	$(1.30)^2 = 1.6900$	$(1.6900)(7) = 11.83$
$4 - 1.7 = 2.30$	$(2.30)^2 = 5.2900$	$(5.2900)(5) = 26.45$
$5 - 1.7 = 3.30$	$(3.30)^2 = 10.9800$	$(10.9800)(1) = 10.89$
$6 - 1.7 = 4.30$	$(4.30)^2 = 18.4900$	$(18.4900)(1) = 18.49$
		$\sum(x - \bar{x})^2 f = 106.5$

d. $s = \sqrt{\dfrac{\sum(x - \bar{x})^2 f}{(n - 1)}} = \sqrt{\dfrac{106.5}{49}} = \sqrt{2.17} \approx 1.5$

10a.

Class	x	f	xf
0–99	49.5	380	$(49.5)(380) = 18,810$
100–199	149.5	230	$(149.5)(230) = 34,385$
200–299	249.5	210	$(249.5)(210) = 52,395$
300–399	349.5	50	$(349.5)(50) = 17,475$
400–499	449.5	60	$(449.5)(60) = 26,970$
500+	650.0	70	$(650.0)(70) = 45,500$
		$n = 1000$	$\sum xf = 195,535$

b. $\bar{x} = \dfrac{\sum xf}{n} = \dfrac{195,535}{1000} \approx 195.5$

c.

$x - \bar{x}$	$(x - \bar{x})^2$	$(x - \bar{x})^2 \cdot f$
$49.5 - 195.5 = -146$	$(-146)^2 = 21,316$	$(21,316)(380) = 8,100,080$
$149.5 - 195.5 = -46$	$(-46)^2 = 2116$	$(2116)(230) = 486,680$
$249.5 - 195.5 = 54$	$(54)^2 = 2916$	$(2916)(210) = 612,360$
$349.5 - 195.5 = 154$	$(154)^2 = 23,716$	$(23,716)(50) = 1,185,800$
$449.5 - 195.5 = 254$	$(254)^2 = 64,516$	$(64,516)(60) = 3,870,960$
$650 - 195.5 = 454.5$	$(454.5)^2 = 206,570.25$	$(206,570.25)(70) = 14,459,917.5$
		$\sum(x - \bar{x})^2 f = 28,715,797.5$

d. $s = \sqrt{\dfrac{\sum(x - \bar{x})^2 f}{n - 1}} = \sqrt{\dfrac{28,715,797.5}{999}} = \sqrt{28,744.542} \approx 169.5$

2.4 EXERCISE SOLUTIONS

1. Range = Max − Min = $12 - 4 = 8$

$\mu = \dfrac{\sum x}{N} = \dfrac{79}{10} = 7.9$

x	$x - \mu$	$(x - \mu)^2$
12	$12 - 7.9 = 4.1$	$(4.1)^2 = 16.81$
9	$9 - 7.9 = 1.1$	$(1.1)^2 = 1.21$
7	$7 - 7.9 = -0.9$	$(-0.9)^2 = 0.81$
5	$5 - 7.9 = -2.9$	$(-2.9)^2 = 8.41$
7	$7 - 7.9 = -0.9$	$(-0.9)^2 = 0.81$
8	$8 - 7.9 = 0.1$	$(0.1)^2 = 0.01$
10	$10 - 7.9 = 2.1$	$(2.1)^2 = 4.41$
4	$4 - 7.9 = -3.9$	$(-3.9)^2 = 15.21$
11	$11 - 7.9 = 3.1$	$(3.1)^2 = 9.61$
6	$6 - 7.9 = -1.9$	$(-1.9)^2 = 3.61$
$\sum x = 79$	$\sum(x - \mu) = 0$	$\sum(x - \mu)^2 = 60.9$

$$\sigma^2 = \frac{\Sigma(x - \mu)^2}{N} = \frac{60.9}{10} = 6.09 \approx 6.1$$

$$\sigma = \sqrt{\frac{\Sigma(x - \mu)^2}{N}} = \sqrt{6.09} \approx 2.5$$

3. Range = Max − Min = 18 − 6 = 12

$$\bar{x} = \frac{\Sigma x}{n} = \frac{107}{9} = 11.9$$

x	$(x - \bar{x})$	$(x - \bar{x})^2$
17	17 − 11.9 = 5.1	$(5.1)^2 = 26.01$
8	8 − 11.9 = −3.9	$(-3.9)^2 = 15.21$
13	13 − 11.9 = 1.1	$(1.1)^2 = 1.21$
18	18 − 11.9 = 6.1	$(6.1)^2 = 37.21$
15	15 − 11.9 = 3.1	$(3.1)^2 = 9.61$
9	9 − 11.9 = −2.9	$(-2.9)^2 = 8.41$
10	10 − 11.9 = −1.9	$(-1.9)^2 = 3.61$
11	11 − 11.9 = −0.9	$(-0.9)^2 = 0.81$
6	6 − 11.9 = −5.9	$(-5.9)^2 = 34.81$
$\Sigma x = 107$	$\Sigma(x - \bar{x}) = 0$	$\Sigma(x - \bar{x})^2 = 136.89$

$$s^2 = \frac{\Sigma(x - \bar{x})^2}{n - 1} = \frac{136.89}{9 - 1} = 17.1$$

$$s = \sqrt{\frac{\Sigma(x - \bar{x})^2}{n - 1}} = \sqrt{17.1} \approx 4.1$$

5. Range = Max − Min = 96 − 23 = 73

7. The range is the difference between the maximum and minimum values of a data set. The advantage of the range is that it is easy to calculate. The disadvantage is that it uses only two entries from the data set.

9. The units of variance are squared. Its units are meaningless. (Ex: dollars²)

11. (a) Range = Max − Min = 45.6 − 21.3 = 24.3

    (b) Range = Max − Min = 65.6 − 21.3 = 44.3

    (c) The range has increased substantially.

13. Graph (a) has a standard deviation of 24 and graph (b) has a standard deviation of 16 because graph (a) has more variability.

15. When calculating the population standard deviation, you divide the sum of the squared deviations by N, then take the square root of that value. When calculating the sample standard deviation, you divide the sum of the squared deviations by $n - 1$, then take the square root of that value.

17. Company B. Due to the larger standard deviation in salaries for company B, it would be more likely to be offered a salary of $33,000.

19. (a) Los Angeles: range = Max − Min = 35.9 − 18.3 = 17.6

x	$(x - \bar{x})$	$(x - \bar{x})^2$
20.2	−6.06	36.67
26.1	−0.16	0.02
20.9	−5.36	28.68
32.1	5.84	34.16
35.9	9.64	93.02
23.0	−3.64	10.60
28.2	1.94	3.78
31.6	5.34	28.56
18.3	−7.96	63.29
$\Sigma x = 236.3$		$\Sigma(x - \bar{x})^2 = 298.78$

$$\bar{x} = \frac{\Sigma x}{n} = \frac{236.3}{9} = 26.26$$

$$s^2 = \frac{\Sigma(x - \bar{x})^2}{(n - 1)} = \frac{298.78}{8} \approx 37.35$$

$$s = \sqrt{s^2} \approx 6.11$$

Long Beach: range = Max − Min = 26.9 − 18.2 = 8.7

x	$(x - \bar{x})$	$(x - \bar{x})^2$
20.9	−1.98	3.91
18.2	−4.68	21.88
20.8	−2.08	4.32
21.1	−1.78	3.16
26.5	3.62	13.12
26.9	4.02	16.18
24.2	1.32	1.75
25.1	2.22	4.94
22.2	−0.68	0.46
$\Sigma x = 205.9$		$\Sigma(x - \bar{x})^2 = 69.72$

$$\bar{x} = \frac{\Sigma x}{n} = \frac{205.9}{9} = 22.88$$

$$s^2 = \frac{\Sigma(x - \bar{x})^2}{(n - 1)} = \frac{69.72}{8} \approx 8.71$$

$$s = \sqrt{s^2} \approx 2.95$$

(b) It appears from the data that the annual salaries in Los Angeles are more variable than the salaries in Long Beach.

21. (a) Male: range = Max − Min = 1328 − 923 = 405

x	$(x - \bar{x})$	$(x - \bar{x})^2$
1059	−51.13	2,613.77
1328	217.8	47,469.52
1175	64.88	4,208.77
1123	12.88	165.77
923	−187.13	35,015.77
1017	−93.13	8,672.77
1214	103.88	10,790.02
1042	−68.13	4,641.02
$\Sigma x = 8881$		$\Sigma(x - \bar{x})^2 = 113{,}576.88$

$$\bar{x} = \frac{\Sigma x}{n} = \frac{8881}{8} = 1110.13$$

$$s^2 = \frac{\Sigma(x - \bar{x})^2}{(n - 1)} = \frac{113{,}576.9}{7} \approx 16{,}225.3$$

$$s = \sqrt{s^2} \approx 127.4$$

Female: range = Max − Min = 1393 − 841 = 552

x	$(x - \bar{x})$	$(x - \bar{x})^2$
1226	92.50	8,556.25
965	−168.50	28,392.25
841	−292.50	85,556.25
1053	−80.50	6,480.25
1056	−77.50	6,006.25
1393	259.50	67,340.25
1312	178.50	31,862.25
1222	88.50	7,832.25
$\Sigma x = 9068$		$\Sigma(x - \bar{x})^2 = 242{,}026.00$

$$\bar{x} = \frac{\Sigma x}{n} = \frac{9068}{8} = 1133.50$$

$$s^2 = \frac{\Sigma(x - \bar{x})^2}{(n - 1)} = \frac{242{,}026}{7} \approx 34{,}575.1$$

$$s = \sqrt{s^2} \approx 185.9$$

(b) It appears from the data, the SAT scores for females are more variable than the SAT scores for males.

23. (a) Greatest sample standard deviation: (ii)

Data set (ii) has more entries that are farther away from the mean.

Least sample standard deviation: (iii)

Data set (iii) has more entries that are close to the mean.

(b) The three data sets have the same mean but have different standard deviations.

25. (a) Greatest sample standard deviation: (ii)

Data set (ii) has more entries that are farther away from the mean.

Least sample standard deviation: (iii)

Data set (iii) has more entries that are close to the mean.

(b) The three data sets have the same mean, median, and mode, but have different standard deviations.

27. Similarity: Both estimate proportions of the data contained within k standard deviations of the mean.

Difference: The Empirical Rule assumes the distribution is bell-shaped, Chebychev's Theorem makes no such assumption.

29. $(1300, 1700) \rightarrow (1500 - 1(200), 1500 + 1(200)) \rightarrow (\bar{x} - s, \bar{x} + s)$

68% of the farms value between \$1300 and \$1700 per acre.

31. (a) $n = 75$

$68\%(75) = (0.68)(75) = 51$ farm values will be between \$1300 and \$1700 per acre.

(b) $n = 25$

$68\%(25) = (0.68)(25) = 17$ of these farm values will be between \$1300 and \$1700 per acre.

33. $\bar{x} = 1500$ {1000, 2000} are outliers. They are more than 2 standard deviations from the mean (1100, 1900).
$s = 200$

35. $(\bar{x} - 2s, \bar{x} + 2s) \rightarrow (1.14, 5.5)$ are 2 standard deviations from the mean.

$1 - \dfrac{1}{k^2} = 1 - \dfrac{1}{(2)^2} = 1 - \dfrac{1}{4} = 0.75 \Rightarrow$ At least 75% of the eruption times lie between 1.14 and 5.5 minutes.

If $n = 32$, at least $(0.75)(32) = 24$ eruptions will lie between 1.14 and 5.5 minutes.

37.

x	f	xf	$x - \bar{x}$	$(x - \bar{x})^2$	$(x - \bar{x})^2 f$
0	5	0	−2.08	4.31	21.53
1	11	11	−1.08	1.16	12.71
2	7	14	−0.08	0.01	0.04
3	10	30	0.93	0.86	8.56
4	7	28	1.93	3.71	25.94
	$n = 40$	$\sum xf = 83$			$\sum (x - \bar{x})^2 f = 68.78$

$$\bar{x} = \frac{\sum x}{n} = \frac{83}{40} \approx 2.1$$

$$s = \sqrt{\frac{\sum (x - \bar{x})^2 f}{n - 1}} = \sqrt{\frac{68.78}{39}} = \sqrt{1.76} \approx 1.3$$

39. Class width $= \dfrac{\text{Max} - \text{Min}}{5} = \dfrac{14 - 2}{5} = \dfrac{12}{5} = 2.4 \Rightarrow 3$

Class	Midpoint, x	f	xf
2–4	3	4	12
5–7	6	8	48
8–10	9	15	135
11–13	12	4	48
14–16	15	1	15
		$N = 32$	$\sum xf = 258$

$$\mu = \frac{\sum xf}{N} = \frac{258}{32} \approx 8.1$$

$x - \mu$	$(x - \mu)^2$	$(x - \mu)^2 f$
−5.1	26.01	104.04
−2.1	4.41	35.28
0.9	0.81	12.15
3.9	15.21	50.84
6.9	47.61	47.61
		$\sum (x - \mu)^2 f = 249.92$

$$\sigma = \sqrt{\frac{\sum (x - \mu)^2}{N}} = \sqrt{\frac{249.92}{32}} \approx 2.8$$

41.

Midpoint, x	f	xf
70.5	1	70.5
92.5	12	1110.0
114.5	25	2862.5
136.5	10	1365.0
158.5	2	317.0
	$n = 50$	$\sum xf = 5725$

$$\bar{x} = \frac{\sum xf}{n} = \frac{5725}{50} = 114.5$$

$x - \bar{x}$	$(x - \bar{x})^2$	$(x - \bar{x})^2 f$
−44	1936	1936
−22	484	5808
0	0	0
22	484	4840
44	1936	3872
		$\sum (x - \bar{x})^2 f = 16,456$

$$s = \sqrt{\frac{\sum (x - \bar{x})^2}{n - 1}} = \sqrt{\frac{16,456}{49}} = \sqrt{335.83} \approx 18.33$$

43.

Class	Midpoint, x	f	xf
0–4	2.0	20.3	40.60
5–13	9.0	35.5	319.50
14–17	15.5	16.5	255.75
18–24	21.0	30.4	638.40
25–34	29.5	39.4	1162.30
35–44	39.5	39.0	1540.50
45–64	54.5	80.8	4403.60
65+	70.0	40.4	2828.00
		$n = 302.3$	$\sum xf = 11{,}188.65$

$$\bar{x} = \frac{\sum xf}{n} = \frac{11{,}188.65}{302.3} \approx 37.01$$

$x - \bar{x}$	$(x - \bar{x})^2$	$(x - \bar{x})^2 f$
−35.01	1225.70	24,881.77
−28.01	784.56	27,851.88
−21.51	462.68	7634.22
−16.01	256.32	7792.13
−7.51	56.40	2222.16
2.49	6.20	241.80
17.49	305.70	24,716.72
32.99	1088.34	43,968.94
		$\sum (x - \bar{x})^2 f = 139{,}309.56$

$$s = \sqrt{\frac{\sum (x - \bar{x})^2}{n - 1}} = \sqrt{\frac{139{,}309.56}{301.3}} = \sqrt{462.36} \approx 21.50$$

45. $CV_{\text{heights}} = \dfrac{\sigma}{\mu} \cdot 100\% = \dfrac{3.44}{72.75} \cdot 100 \approx 4.7$

$CV_{\text{weights}} = \dfrac{\sigma}{\mu} \cdot 100\% = \dfrac{18.47}{187.83} \cdot 100 \approx 9.8$

It appears that weight is more variable than height.

47. (a) $\bar{x} \approx 41.5 \qquad s \approx 5.3$

(b) $\bar{x} \approx 43.6 \qquad s \approx 5.6$

(c) $\bar{x} \approx 3.5 \qquad s \approx 0.4$

(d) By multiplying each entry by a constant k, the new sample mean is $k \cdot \bar{x}$ and the new sample standard deviation is $k \cdot s$.

49. (a) Male SAT Scores: $\bar{x} = 1110.125$ ⠀⠀⠀⠀ Female SAT Score: $\bar{x} = 1133.5$

x	$\lvert x - \bar{x} \rvert$
1059	51.125
1328	217.88
1175	64.875
1123	12.875
923	187.13
1017	93.125
1214	103.88
1042	68.125
	$\sum \lvert x - \bar{x} \rvert = 799$

x	$\lvert x - \bar{x} \rvert$
1226	92.5
965	168.5
841	292.5
1053	80.5
1056	77.5
1393	259.5
1312	178.5
1222	88.5
	$\sum \lvert x - \bar{x} \rvert = 1238$

$\sum \lvert x - \bar{x} \rvert = 799 \Rightarrow \dfrac{\sum \lvert x - \bar{x} \rvert}{n} = \dfrac{799}{8} = 99.9$

$s = 127.4$

$\sum \lvert x - \bar{x} \rvert = 799 \Rightarrow \dfrac{\sum \lvert x - \bar{x} \rvert}{n} = \dfrac{1238}{8} = 154.8$

$s = 185.9$

(b) Public Teachers: $\bar{x} = 37.375$ Private Teachers: $\bar{x} = 19.538$

x	$|x - \bar{x}|$
38.6	1.225
38.1	0.725
38.7	1.325
36.8	0.575
34.8	2.575
35.9	1.475
39.9	2.525
36.2	1.175
	$\Sigma|x - \bar{x}| = 11.6$

x	$|x - \bar{x}|$
21.8	2.262
18.4	1.138
20.3	0.762
17.6	1.938
19.7	0.162
18.3	1.238
19.4	0.138
20.8	1.262
	$\Sigma|x - \bar{x}| = 8.9$

$$\Sigma|x - \bar{x}| = 11.6 \Rightarrow \frac{\Sigma|x - \bar{x}|}{n} = \frac{11.6}{8} = 1.45$$

$$s = 1.72$$

$$\Sigma|x - \bar{x}| = 8.9 \Rightarrow \frac{\Sigma|x - \bar{x}|}{n} = \frac{8.9}{8} = 1.11$$

$$s = 1.41$$

51. (a) $P = \dfrac{3(\bar{x} - \text{median})}{s} = \dfrac{3(17 - 19)}{2.3} \approx -2.61$; skewed left

(b) $P = \dfrac{3(\bar{x} - \text{median})}{s} = \dfrac{3(32 - 25)}{5.1} \approx 4.12$; skewed right

2.5   MEASURES OF POSITION

2.5 Try It Yourself Solutions

1a. 15, 15, 15, 17, 18, 18, 20, 21, 21, 21, 21, 22, 22, 23, 24, 24, 25, 25, 26, 26, 27, 27, 27, 28, 28, 28, 29, 29, 29, 30, 30, 31, 31, 31, 31, 31, 32, 32, 32, 32, 32, 32, 32, 33, 33, 33, 34, 34, 34, 34, 35, 35, 35, 35, 35, 36, 36, 37, 37, 37, 38, 38, 38, 39, 39, 39, 40, 40, 40, 41, 41, 42, 42, 43, 43, 43, 44, 44, 44, 44, 45, 45, 45, 45, 46, 46, 46, 46, 47, 47, 48, 48, 48, 48, 49, 50, 50, 52, 53, 53, 54, 54, 55, 55, 55, 55, 57, 59, 59, 60, 61, 61, 61, 63, 63, 65, 68, 89

b. $Q_2 = 37$

c. $Q_1 = 30$ $Q_3 = 47$

2a. (Enter the data)

b. $Q_1 = 17$ $Q_2 = 23$ $Q_3 = 28.5$

c. One quarter of the tuition costs is $17,000 or less, one half is $23,000 or less, and three quarters is $28,500 or less.

3a. $Q_1 = 30$ $Q_3 = 47$

b. IQR $= Q_3 - Q_1 = 47 - 30 = 17$

c. The touchdowns in the middle half of the data set vary by 17 years.

4a. Min = 15 $Q_1 = 30$ $Q_2 = 37$

   $Q_3 = 47$ Max = 89

bc.

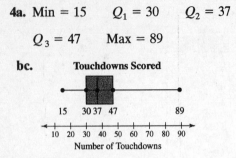

d. It appears that half of the teams scored between 30 and 47 touchdowns.

5a. 50th percentile

b. 50% of the teams scored 40 or fewer touchdowns.

6a. $\mu = 70, \sigma = 8$

b. $x = 60$: $z = \dfrac{x - \mu}{\sigma} = \dfrac{60 - 70}{8} = -1.25$

   $x = 71$: $z = \dfrac{x - \mu}{\sigma} = \dfrac{71 - 70}{8} = 0.125$

   $x = 92$: $z = \dfrac{x - \mu}{\sigma} = \dfrac{92 - 70}{8} = 2.75$

c. From the z-score, the utility bill of $60 is 1.25 standard deviations below the mean, the bill of $71 is 0.125 standard deviation above the mean, and the bill of $92 is 2.75 standard deviations above the mean.

7a. Best supporting actor: $\mu = 50.1, \sigma = 13.9$

   Best supporting actress: $\mu = 39.7, \sigma = 1.4$

b. Alan Arkin: $x = 72 \Rightarrow z = \dfrac{x - \mu}{\sigma} = \dfrac{72 - 50.1}{13.9} = 1.58$

   Jennifer Hudson: $x = 25 \Rightarrow z = \dfrac{x - \mu}{\sigma} = \dfrac{25 - 39.7}{14} = -1.05$

c. Alan Arkin's age is 1.58 standard deviations above the mean of the best support actors. Jennifer Hudson's age is 1.05 standard deviations below the mean of the best supporting actresses. Neither actor's age is unusual.

2.5 EXERCISE SOLUTIONS

1. (a)

        lower half   upper half

1 2 2 4 4 5 5 5 6 6 6 7 7 7 7 8 8 8 9 9

               Q_1 Q_2 Q_3

        $Q_1 = 4.5$ $Q_2 = 6$ $Q_3 = 7.5$

(b)

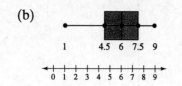

1   4.5 6 7.5 9

0 1 2 3 4 5 6 7 8 9

3. The soccer team scored fewer points per game than 75% of the teams in the league.

5. The student scored higher than 78% of the students who took the actuarial exam.

7. True

9. False. The 50th percentile is equivalent to Q_2.

11. (a) Min = 10   (b) Max = 20

    (c) $Q_1 = 13$ (d) $Q_2 = 15$

    (e) $Q_3 = 17$ (f) IQR $= Q_3 - Q_1 = 17 - 13 = 4$

13. (a) Min = 900   (b) Max = 2100

    (c) $Q_1 = 1250$ (d) $Q_2 = 1500$

    (e) $Q_3 = 1950$ (f) IQR $= Q_3 - Q_1 = 1950 - 1250 = 700$

15. (a) Min = -1.9 (b) Max = 2.1

    (c) $Q_1 = -0.5$ (d) $Q_2 = 0.1$

    (e) $Q_3 = 0.7$ (f) IQR $= Q_3 - Q_1 = 0.7 - (-0.5) = 1.2$

17. None. The data are not skewed or symmetric.

19. Skewed left. Most of the data lie to the left.

21. $Q_1 = B$, $Q_2 = A$, $Q_3 = C$

    25% of the entries are below B, 50% are below A, and 75% are below C.

23. (a) $Q_1 = 2$, $Q_2 = 4$, $Q_3 = 5$

    (b)   **Watching Television**

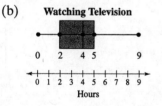

0   2   4 5   9

0 1 2 3 4 5 6 7 8 9

Hours

25. (a) $Q_1 = 3$, $Q_2 = 3.85$, $Q_3 = 5.28$

    (b)   **Airline Distances**

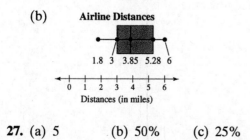

1.8 3 3.85 5.28 6

0   1   2   3   4   5   6

Distances (in miles)

27. (a) 5   (b) 50%   (c) 25%

29. $A \Rightarrow z = -1.43$

$B \Rightarrow z = 0$

$C \Rightarrow z = 2.14$

The z-score 2.14 is unusual because it is so large.

31. (a) Statistics:

$$x = 73 \Rightarrow z = \frac{x - \mu}{\sigma} = \frac{73 - 63}{7} \approx 1.43$$

Biology:

$$x = 26 \Rightarrow z = \frac{x - \mu}{\sigma} = \frac{26 - 23}{3.9} \approx 0.77$$

(b) The student had a better score on the statistics test.

33. (a) Statistics:

$$x = 78 \Rightarrow z = \frac{x - \mu}{\sigma} = \frac{78 - 63}{7} \approx 2.14$$

Biology:

$$x = 29 \Rightarrow z = \frac{x - \mu}{\sigma} = \frac{29 - 23}{3.9} \approx 1.54$$

(b) The student had a better score on the statistics test.

35. (a) $x = 34{,}000 \Rightarrow z = \dfrac{x - \mu}{\sigma} = \dfrac{34{,}000 - 35{,}000}{2{,}250} \approx -0.44$

$x = 37{,}000 \Rightarrow z = \dfrac{x - \mu}{\sigma} = \dfrac{37{,}000 - 35{,}000}{2{,}250} \approx 0.89$

$x = 31{,}000 \Rightarrow z = \dfrac{x - \mu}{\sigma} = \dfrac{31{,}000 - 35{,}000}{2{,}250} \approx -1.78$

None of the selected tires have unusual life spans.

(b) $x = 30{,}500 \Rightarrow z = \dfrac{x - \mu}{\sigma} = \dfrac{30{,}500 - 35{,}000}{2{,}250} = -2 \Rightarrow 2.5\text{th percentile}$

$x = 37{,}250 \Rightarrow z = \dfrac{x - \mu}{\sigma} = \dfrac{37{,}250 - 35{,}000}{2{,}250} = 1 \Rightarrow 84\text{th percentile}$

$x = 35{,}000 \Rightarrow z = \dfrac{x - \mu}{\sigma} = \dfrac{35{,}000 - 35{,}000}{2{,}250} = 0 \Rightarrow 50\text{th percentile}$

37. 68.5 inches

40% of the heights are below 68.5 inches.

39. $x = 74$: $z = \dfrac{x - \mu}{\sigma} = \dfrac{74 - 69.6}{3.0} \approx 1.47$

$x = 62$: $z = \dfrac{x - \mu}{\sigma} = \dfrac{62 - 69.6}{3.0} \approx -2.53$

$x = 80$: $z = \dfrac{x - \mu}{\sigma} = \dfrac{80 - 69.6}{3.0} \approx 3.47$

The height of 62 inches is unusual due to a rather small z-score. The height of 80 inches is very unusual due to a rather large z-score.

41. $x = 71.1$: $z = \dfrac{x - \mu}{\sigma} = \dfrac{71.1 - 69.6}{3.0} \approx 0.5$

Approximately the 70th percentile.

43. (a) 27 28 31 32 32 33 35 36 36 36 36 37 38 39 39 40 40 40 41 41

41 42 42 42 42 42 42 43 43 43 44 44 45 45 46 47 47 47 47 47

48 48 48 48 48 | 49 49 49 49 49 49 50 50 51 51 51 51 51 51 52

52 52 53 53 54 | 54 54 54 54 54 | 54 54 55 56 56 56 57 57 57 59

59 59 60 60 60 | 61 61 61 62 62 | 63 63 63 63 64 | 65 67 68 74 82

Q_1 Q_2 Q_3

$Q_1 = 42$, $Q_2 = 49$, $Q_3 = 56$

(b) **Ages of Executives**

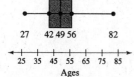

27   42 49 56   82

25 35 45 55 65 75 85
Ages

(c) Half of the ages are between 42 and 56 years.

(d) About 49 years old ($x = 49.62$ and $Q_2 = 49.00$), because half of the executives are older and half are younger.

(e) The age groups 20–29, 70–79, and 80–89 would all be considered unusual because they lie more than two standard deviations from the mean.

45. 22   23   24   32   33   34   36   38   39   40   41   47

$Q_1 = 28$ Q_2 $Q_3 = 39.5$

Midquartile $= \dfrac{Q_1 + Q_3}{2} = \dfrac{28 + 39.5}{2} = 33.75$

47. 13.4 15.2 15.6 16.7 17.2 18.7 19.7 19.8 19.8 20.8 21.4 22.9 28.7 30.1 31.9

Q_1 Q_2 Q_3

Midquartile $= \dfrac{Q_1 + Q_3}{2} = \dfrac{16.7 + 22.9}{2} = 19.8$

49.

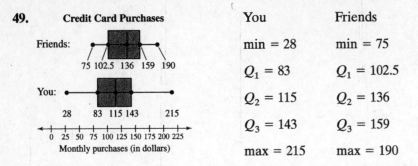

Credit Card Purchases

Friends: 75 102.5 136 159 190

You: 28 83 115 143 215

0 25 50 75 100 125 150 175 200 225

Monthly purchases (in dollars)

You	Friends
min = 28	min = 75
$Q_1 = 83$	$Q_1 = 102.5$
$Q_2 = 115$	$Q_2 = 136$
$Q_3 = 143$	$Q_3 = 159$
max = 215	max = 190

Your distribution is symmetric and your friend's distribution is uniform.

51. $\text{Percentile} = \dfrac{\text{Number of data values less than } x}{\text{Total number of data values}} \cdot 100 = \dfrac{75}{80} \cdot 100 \approx 94\text{th percentile}$

CHAPTER 2 REVIEW EXERCISE SOLUTIONS

1.

Class	Midpoint	Boundaries	Frequency, f	Relative frequency	Cumulative frequency
20–23	21.5	19.5–23.5	1	0.05	1
24–27	25.5	23.5–27.5	2	0.10	3
28–31	29.5	27.5–31.5	6	0.30	9
32–35	33.5	31.5–35.5	7	0.35	16
36–39	37.5	35.5–39.5	4	0.20	20
			$\Sigma f = 20$	$\Sigma \dfrac{f}{n} = 1$	

3.

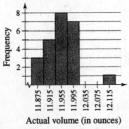

Liquid Volume 12-oz Cans

Frequency

11.875 11.915 11.955 11.995 12.035 12.075 12.115

Actual volume (in ounces)

5.

Class	Midpoint, x	Frequency, f	Cumulative frequency
79–93	86	9	9
94–108	101	12	21
109–123	116	5	26
124–138	131	3	29
139–153	146	2	31
154–168	161	1	32
		$\Sigma f = 32$	

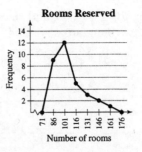

Rooms Reserved

Frequency

71 86 101 116 131 146 161 176

Number of rooms

7.

```
1 | 3 7 8 9
2 | 0 1 2 3 3 3 4 4 5 5 5 7 8 8 9
3 | 1 1 2 3 4 5 7 8
4 | 3 4 7
5 | 1
```

9.

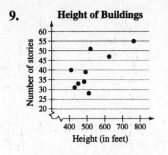

Height of Buildings

It appears as height increases, the number of stories increases.

11.

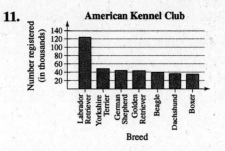

American Kennel Club

13. $\bar{x} = 9.1$

median $= 8.5$

mode $= 7$

15.

Midpoint, x	Frequency, f	xf
21.5	1	21.5
25.5	2	51.0
29.5	6	177.0
33.5	7	234.5
37.5	4	150.0
	$n = 20$	$\Sigma xf = 634$

$$\bar{x} = \frac{\Sigma xf}{n} = \frac{634}{20} = 31.7$$

17. $\bar{x} = \dfrac{\Sigma xw}{w} = \dfrac{(78)(0.15) + (72)(0.15) + (86)(0.15) + (91)(0.15) + (87)(0.15) + (80)(0.25)}{0.15 + 0.15 + 0.15 + 0.15 + 0.15 + 0.25}$

$\qquad = \dfrac{82.1}{1} = 82.1$

19. Skewed   **21.** Skewed left   **23.** Median

25. Range $=$ Max $-$ Min $= 8.26 - 5.46 = 2.8$

27. $\mu = \dfrac{\Sigma x}{N} = \dfrac{96}{14} = 6.9$

$\sigma = \sqrt{\dfrac{\Sigma(x - \mu)^2}{N}} = \sqrt{\dfrac{(4 - 6.9)^2 + (2 - 6.9)^2 + \cdots + (3 - 6.9)^2 + (3 - 6.9)^2}{12}}$

$\quad = \sqrt{\dfrac{295.7}{14}} \approx \sqrt{21.12} \approx 4.6$

29. $\bar{x} = \dfrac{\Sigma x}{n} = \dfrac{36{,}801}{15} = 2453.4$

$s = \sqrt{\dfrac{\Sigma(x - \bar{x})^2}{n - 1}} = \sqrt{\dfrac{(2445 - 2453.4)^2 + \cdots + (2.377 - 2453.4)^2}{14}}$

$\quad = \sqrt{\dfrac{1{,}311{,}783.6}{14}} \approx \sqrt{93{,}698.8} \approx 306.1$

31. 99.7% of the distribution lies within 3 standard deviations of the mean.

$$\mu + 3\sigma = 49 + (3)(2.50) = 41.5$$

$$\mu - 3\sigma = 49 - (3)(2.50) = 56.5$$

99.7% of the distribution lies between $41.50 and $56.50.

33. $n = 40$ $\mu = 36$ $\sigma = 8$

$$(20, 52) \rightarrow (36 - 2(8), 36 + 2(8)) \Rightarrow (\mu - 2\sigma, \mu + 2\sigma) \Rightarrow k = 2$$

$$1 = \frac{1}{k^2} = 1 - \frac{1}{(2)^2} = 1 - \frac{1}{4} = 0.75$$

At least $(40)(0.75) = 30$ customers have a mean sale between $20 and $52.

35. $\bar{x} = \dfrac{\Sigma xf}{n} = \dfrac{99}{40} \approx 2.5$

$$s = \sqrt{\frac{\Sigma (x - \bar{x})^2 f}{n - 1}} = \sqrt{\frac{(0 - 1.24)^2(1) + (1 - 1.24)^2(8) + \cdots + (5 - 1.24)^2(3)}{39}}$$

$$= \sqrt{\frac{59.975}{39}} \approx 1.2$$

37. $Q_1 = 56$ inches

39. $\text{IQR} = Q_3 - Q_1 = 68 - 56 = 12$ inches

41. $\text{IQR} = Q_3 - Q_1 = 33 - 29 = 4$

43. 23% of the students scored higher than 68.

45. $x = 213 \Rightarrow z = \dfrac{x - \mu}{\sigma} = \dfrac{213 - 186}{18} \approx 1.5$

This player is not unusual.

47. $x = 178 \Rightarrow z = \dfrac{x - \mu}{\sigma} = \dfrac{178 - 186}{18} = -0.44$

This player is not unusual.

CHAPTER 2 QUIZ SOLUTIONS

1. (a)

Class limits	Midpoint	Class boundaries	Frequency, f	Relative frequency	Cumulative frequency
101–112	106.5	100.5–112.5	3	0.12	3
113–124	118.5	112.5–124.5	11	0.44	14
125–136	130.5	124.5–136.5	7	0.28	21
137–148	142.5	136.5–148.5	2	0.08	23
149–160	154.5	148.5–160.5	2	0.08	25

(b) Frequency Histogram and Polygon

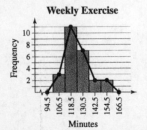

(c) Relative Frequency Histogram

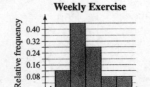

(d) Skewed

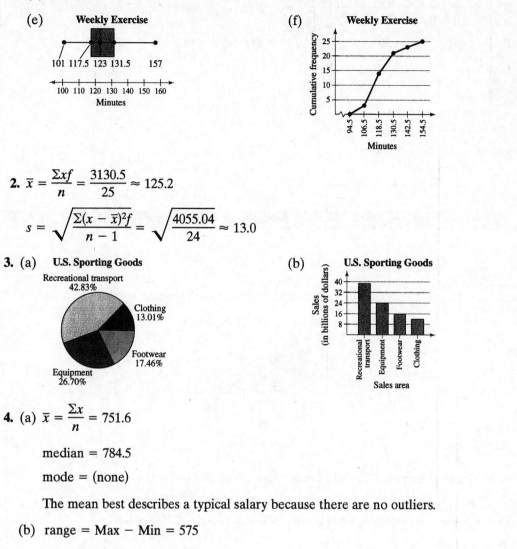

(e) **Weekly Exercise**

101 117.5 123 131.5   157

100 110 120 130 140 150 160
Minutes

(f) **Weekly Exercise**

Cumulative frequency

25
20
15
10
5

94.5 106.5 118.5 130.5 142.5 154.5
Minutes

2. $\bar{x} = \dfrac{\Sigma xf}{n} = \dfrac{3130.5}{25} \approx 125.2$

$s = \sqrt{\dfrac{\Sigma(x - \bar{x})^2 f}{n - 1}} = \sqrt{\dfrac{4055.04}{24}} \approx 13.0$

3. (a) **U.S. Sporting Goods**

Recreational transport
42.83%

Clothing
13.01%

Footwear
17.46%

Equipment
26.70%

(b) **U.S. Sporting Goods**

Sales
(in billions of dollars)

40
32
24
16
8

Recreational transport | Equipment | Footwear | Clothing

Sales area

4. (a) $\bar{x} = \dfrac{\Sigma x}{n} = 751.6$

median = 784.5

mode = (none)

The mean best describes a typical salary because there are no outliers.

(b) range = Max − Min = 575

$s^2 = \dfrac{\Sigma(x - \bar{x})^2}{n - 1} = 48{,}135.1$

$s = \sqrt{\dfrac{\Sigma(x - \bar{x})^2}{n - 1}} = 219.4$

5. $\bar{x} - 2s = 155{,}000 - 2 \cdot 15{,}000 = \$125{,}000$

$\bar{x} + 2s = 155{,}000 + 2 \cdot 15{,}000 = \$185{,}000$

95% of the new home prices fall between \$125,000 and \$185,000.

6. (a) $x = 200{,}000$ $z = \dfrac{x - \mu}{\sigma} = \dfrac{200{,}000 - 155{,}000}{15{,}000} = 3.0 \Rightarrow$ unusual price

(b) $x = 55{,}000$ $z = \dfrac{x - \mu}{\sigma} = \dfrac{55{,}000 - 155{,}000}{15{,}000} \approx -6.67 \Rightarrow$ very unusual price

(c) $x = 175{,}000$ $z = \dfrac{x - \mu}{\sigma} = \dfrac{175{,}000 - 155{,}000}{15{,}000} \approx 1.33 \Rightarrow$ not unusual

(d) $x = 122{,}000$ $z = \dfrac{x - \mu}{\sigma} = \dfrac{122{,}000 - 155{,}000}{15{,}000} = -2.2 \Rightarrow$ unusual price

7. (a) $Q_1 = 76$ $Q_2 = 80$ $Q_3 = 88$ (b) $\text{IQR} = Q_3 - Q_1 = 88 - 76 = 12$

(c)

Wins for Each Team

61   76   80   88   97

Number of wins

CUMULATIVE REVIEW FOR CHAPTERS 1 AND 2

1. Systematic sampling

2. Simple Random Sampling. However, all U.S. adults may not have a telephone.

3.
Personal Illness	35%
Family Issues	24%
Personal Needs	18%
Stress	12%
Entitlement Mentality	11%

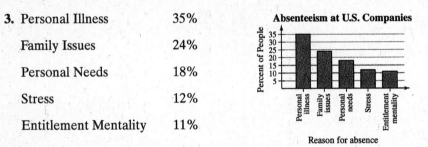

Absenteeism at U.S. Companies

4. $42,500 is a parameter because it is describing the average salary of all 43 employees in a company.

5. 28% is a statistic because it is describing a proportion within a sample of 1000 adults.

6. (a) $\bar{x} = 83{,}500, s = \$1500$

$(80{,}500 \;\; 86{,}500) = 83{,}500 \pm 2(1500) \Rightarrow 2$ standard deviations away from the mean.

Approximately 95% of the electrical engineers will have salaries between $(80{,}500, \; 86{,}500)$.

(b) $40(.95) = 38$

7. Sample, because a survey of 1498 adults was taken.

8. Sample, because a study of 232,606 people was done.

9. Census, because all of the members of the Senate were included.

10. Experiment, because we want to study the effects of removing recess from schools.

11. Quantitative: Ratio

12. Qualitative: Nominal

13. min = 0

$Q_1 = 2$

$Q_2 = 12.5$

$Q_3 = 39$

max = 136

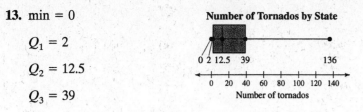

Number of Tornados by State

0 2 12.5 39   136

Number of tornados

14. $\bar{x} = \dfrac{(0.15)(85) + (0.15)(92) + (0.15)(84) + (215)(89) + (0.40)(91)}{0.15 + 0.15 + 0.15 + 0.15 + 0.40} = 88.9$

15. (a) $\bar{x} = 5.49$

median = 5.4

mode = none

Median, because the distribution is not symmetric.

(b) Range = 4.1

$s^2 = 2.34$

$s = 1.53$

The standard deviation of tail lengths of alligators is 1.53 feet.

16. (a) The number of deaths due to heart disease for women is decreasing.

(b) The study was only conducted over the past 5 years and deaths may not decrease in the next year.

17. Class width $= \dfrac{87 - 0}{8} = 10.875 \Rightarrow 11$

Class limits	Class boundaries	Class midpoint	Frequency, f	Relative frequency	Cumulative frequency
0–10	−0.5–10.5	5	8	0.296	8
11–21	10.5–21.5	16	8	0.296	16
22–32	21.5–32.5	27	1	0.037	17
33–43	32.5–43.5	38	1	0.037	18
44–54	43.5–54.5	49	1	0.037	19
55–65	54.5–65.5	60	4	0.148	23
66–76	65.5–76.5	71	0	0.000	23
77–87	76.5–87.5	82	4	0.148	27
			$n = 27$	$\sum \dfrac{f}{n} = 1$	

18. Skewed right

19.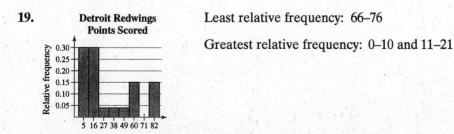

Detroit Redwings Points Scored

Least relative frequency: 66–76

Greatest relative frequency: 0–10 and 11–21

Probability

3.1 Try It Yourself Solutions

1ab. (1)

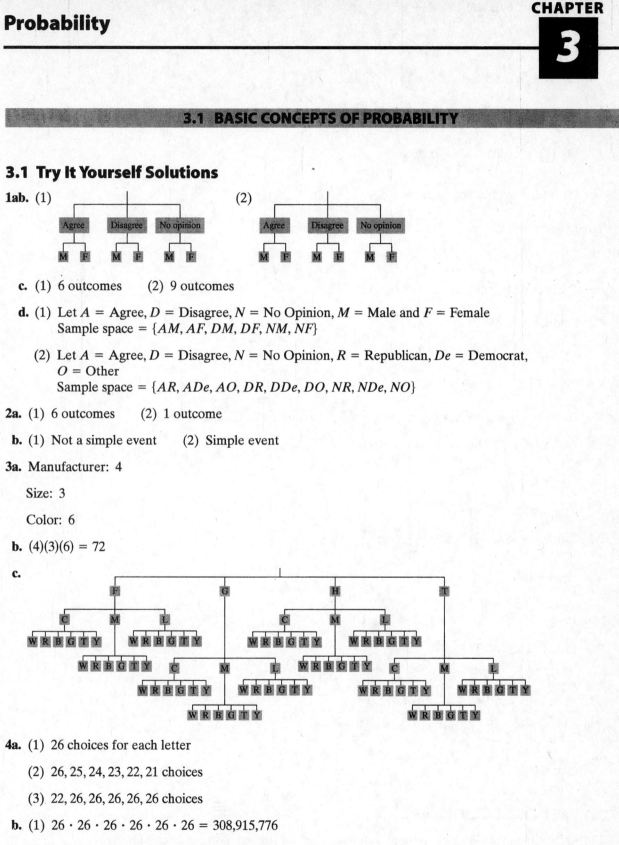

c. (1) 6 outcomes   (2) 9 outcomes

d. (1) Let A = Agree, D = Disagree, N = No Opinion, M = Male and F = Female
Sample space = {AM, AF, DM, DF, NM, NF}

   (2) Let A = Agree, D = Disagree, N = No Opinion, R = Republican, De = Democrat, O = Other
Sample space = {$AR, ADe, AO, DR, DDe, DO, NR, NDe, NO$}

2a. (1) 6 outcomes   (2) 1 outcome

b. (1) Not a simple event   (2) Simple event

3a. Manufacturer: 4

   Size: 3

   Color: 6

b. (4)(3)(6) = 72

c.

4a. (1) 26 choices for each letter

   (2) 26, 25, 24, 23, 22, 21 choices

   (3) 22, 26, 26, 26, 26, 26 choices

b. (1) $26 \cdot 26 \cdot 26 \cdot 26 \cdot 26 \cdot 26 = 308{,}915{,}776$

   (2) $26 \cdot 25 \cdot 24 \cdot 23 \cdot 22 \cdot 21 = 165{,}765{,}600$

   (3) $22 \cdot 26 \cdot 26 \cdot 26 \cdot 26 \cdot 26 = 261{,}390{,}272$

45

5a. (1) 52　　(2) 52　　(3) 52

b. (1) 1　　(2) 13　　(3) 52

c. (1) $P(7 \text{ of diamonds}) = \dfrac{1}{52} \approx 0.0192$

(2) $P(\text{diamond}) = \dfrac{13}{52} = 0.25$

(3) $P(\text{diamond, heart, club, or spade}) = \dfrac{52}{52} = 1$

6a. Event = the next claim processed is fraudulent (Freq = 4)

b. Total Frequency = 100

c. $P(\text{fraudulent claim}) = \dfrac{4}{100} = 0.04$

7a. Frequency = 54

b. Total of the Frequencies = 1000

c. $P(\text{age 15 to 24}) = \dfrac{54}{1000} = 0.054$

8a. Event = salmon successfully passing through a dam on the Columbia River.

b. Estimated from the results of an experiment.

c. Empirical probability

9a. $P(\text{age 45 to 54}) = \dfrac{180}{1000} = 0.18$

b. $P(\text{not age 45 to 54}) = 1 - \dfrac{180}{1000} = \dfrac{820}{1000} = 0.82$

c. $\dfrac{820}{1000}$ or 0.82

10a. 16

b. {T1, T2, T3, T4, T5}

c. $P(\text{tail and less than 6}) = \dfrac{5}{16} = 0.3125$

11a. $10 \cdot 10 \cdot 10 \cdot 10 \cdot 10 \cdot 10 \cdot 10 = 10{,}000{,}000$

b. $\dfrac{1}{10{,}000{,}000}$

3.1 EXERCISE SOLUTIONS

1. (a) Yes, the probability of an event occurring must be contained in the interval [0, 1] or [0%, 100%].

(b) Yes, the probability of an event occurring must be contained in the interval [0, 1] or [0%, 100%].

(c) No, the probability of an event occurring cannot be less than 0.

(d) Yes, the probability of an event occurring must be contained in the interval $[0, 1]$ or $[0\%, 100\%]$.

(e) Yes, the probability of an event occurring must be contained in the interval $[0, 1]$ or $[0\%, 100\%]$.

(f) No, the probability of an event cannot be greater than 1.

3. The fundamental counting principle counts the number of ways that two or more events can occur in sequence.

5. {A, B, C, D, E, F, G, H, I, J, K, L, M, N, O, P, Q, R, S, T, U, V, W, X, Y, Z}; 26

7. 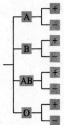 {$(A, +), (B, +), (AB, +), (O, +), (A, -), (B, -), (AB, -), (O, -)$} where $(A, +)$ represents positive Rh-factor with A- blood type and $(A, -)$ represents negative Rh-factor with A- blood type; 8.

9. Simple event because it is an event that consists of a single outcome.

11. Not a simple event because it is an event that consists of more than a single outcome.

king = {king of hearts, king of spades, king of clubs, king of diamonds}

13. $(9)(15) = 135$

15. $(9)(10)(10)(5) = 4500$

17. False. If you roll a six-sided die six times, the probability of rolling an even number at least once is approximately 0.9844.

19. False. A probability of less than 0.05 indicates an unusual event.

21. b   **23.** c

25. Empirical probability because company records were used to calculate the frequency of a washing machine breaking down.

27. $P(\text{less than 1000}) = \dfrac{999}{6296} \approx 0.159$

29. $P(\text{number divisible by 1000}) = \dfrac{6}{6296} \approx 0.000953$

31. $P(A) = \dfrac{1}{24} \approx 0.042$

33. $P(C) = \dfrac{5}{24} \approx 0.208$

35. (a) $10 \cdot 10 \cdot 10 = 1000$ (b) $\dfrac{1}{1000} = 0.001$ (c) $\dfrac{999}{1000} = 0.999$

37. {(SSS), (SSR), (SRS), (SRR), (RSS), (RSR), (RRS), (RRR)}

39. {(SSR), (SRS), (RSS)}

41. Let S = Sunny Day and R = Rainy Day

(a)

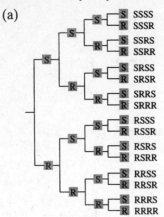

(b) {(SSSS), (SSSR), (SSRS), (SSRR), (SRSS), (SRSR)

(SRRS), (SRRR), (RSSS), (RSSR), (RSRS), (RSRR),

(RRSS), (RRSR), (RRRS), (RRRR)}

(c) {(SSSR), (SSRS), (SRSS), (RSSS)}

43. $P(\text{voted in 2006 Gubernatorial election}) = \dfrac{4,092,652}{8,182,876} \approx 0.500$

45. $P(\text{between 21 and 24}) = \dfrac{8.5}{142.1} \approx 0.060$

47. $P(\text{not between 18 and 20}) = 1 - \dfrac{5.8}{142.1} \approx 1 - 0.041 \approx 0.959$

49. $P(\text{Ph.D.}) = \dfrac{8}{89} \approx 0.090$ **51.** $P(\text{master's}) = \dfrac{21}{89} \approx 0.236$

53. (a) $P(\text{pink}) = \dfrac{2}{4} = 0.5$ (b) $P(\text{red}) = \dfrac{1}{4} = 0.25$ (c) $P(\text{white}) = \dfrac{1}{4} = 0.25$

55. $P(\text{service industry}) = \dfrac{113,409}{144,428} \approx 0.785$

57. $P(\text{not in service industry}) = 1 - P(\text{service industry}) = 1 - 0.785 = 0.215$

59. (a) $P(\text{at least 21}) = \dfrac{87}{118} \approx 0.739$

(b) $P(\text{between 40 and 50 inclusive}) = \dfrac{32}{118} \approx 0.269$

(c) $P(\text{older than 69}) = \dfrac{1}{118} = 0.008$

61. The probability of choosing a tea drinker who does not have a college degree.

63. (a)

Sum	P(sum)	Probability
2	1/36	0.028
3	2/36	0.056
4	3/36	0.083
5	4/36	0.111
6	5/36	0.139
7	6/36	0.167
8	5/36	0.139
9	4/36	0.111
10	3/36	0.083
11	2/36	0.056
12	1/36	0.028

(b) Answers will vary.

(c) The answers in part (a) and (b) will be similar.

65. (a) $P(\text{event will occur}) = \dfrac{4}{9} \approx 0.444$

(b) $P(\text{event will not occur}) = \dfrac{5}{9} \approx 0.556$

67. $39:13 = 3:1$

3.2 CONDITIONAL PROBABILITY AND THE MULTIPLICATION RULE

3.2 Try It Yourself Solutions

1a. (1) 30 and 102   (2) 11 and 50

  b. (1) $P(\text{not have gene}) = \dfrac{30}{102} \approx 0.294$ (2) $P(\text{not have gene}|\text{normal IQ}) = \dfrac{11}{50} = 0.22$

2a. (1) Yes   (2) No

  b. (1) Dependent   (2) Independent

3a. (1) Independent   (2) Dependent

  b. (1) Let $A = \{\text{swimming through 1st dam}\}$

       $B = \{\text{swimming through 2nd dam}\}$

       $P(A \text{ and } B) = P(A) \cdot P(B|A) = (0.85) \cdot (0.85) = 0.723$

    (2) Let $A = \{\text{selecting a heart}\}$

       $B = \{\text{selecting a second heart}\}$

       $P(A \text{ and } B) = P(A) \cdot P(B|A) = \left(\dfrac{13}{52}\right) \cdot \left(\dfrac{12}{51}\right) \approx 0.059$

4a. (1) Find probability of the event   (2) Find probability of the complement of the event

  b. (1) $P(3 \text{ knee surgeries successful}) = (0.90) \cdot (0.90) \cdot (0.90) = 0.729$

    (2) $P(\text{at least one knee surgery successful}) = 1 - P(\text{none are successful})$

                                   $= 1 - (0.10) \cdot (0.10) \cdot (0.10) = 0.999$

5a. (1)(2) $A = \{\text{is female}\}; B = \{\text{works in health field}\}$

  b. (1) $P(A \text{ and } B) = P(A)P(B|A) = (0.65)(0.25)$

    (2) $P(A \text{ and } B') = P(A)(1 - P(B|A)) = (0.65)(0.75)$

  c. (1) 0.1625

    (2) 0.4875

3.2 EXERCISE SOLUTIONS

1. Two events are independent if the occurrence of one of the events does not affect the probability of the occurrence of the other event.

    If $P(B|A) = P(B)$ or $P(A|B) = P(A)$, then Events A and B are independent.

3. False. If two events are independent, $P(A|B) = P(A)$.

5. These events are independent because the outcome of the 1st card drawn does not affect the outcome of the 2nd card drawn.

7. These events are dependent because the sum of the rolls depends on which numbers were rolled first and second.

9. Events: depression, breathing-related sleeping disorder

    Dependent. People with depression are more likely to have a breathing-related sleeping disorder.

11. Events: memory loss, use of Aspartame

    Independent. The use of Aspartame does not cause memory loss.

13. Let A = {have mutated BRCA gene} and B = {develop breast cancer}. Thus

    $P(B) = \frac{1}{8}$, $P(A) = \frac{1}{600}$, and $P(B|A) = \frac{8}{10}$.

    (a) $P(B|A) = \frac{8}{10} = 0.8$

    (b) $P(A \text{ and } B) = P(A) \cdot P(B|A) = \left(\frac{1}{600}\right) \cdot \left(\frac{8}{10}\right) = 0.0013$

    (c) Dependent because $P(B|A) \neq P(B)$.

15. Let A = {own a computer} and B = {summer vacation this year}.

    (a) $P(B') = \frac{45}{146} = 0.308$

    (b) $P(A) = \frac{57}{146} = 0.390$

    (c) $P(B|A) = \frac{46}{57} \approx 0.807$

    (d) $P(A \text{ and } B) = P(A)P(B|A) = \left(\frac{57}{146}\right)\left(\frac{46}{57}\right) = 0.315$

    (e) Dependent, because $P(B|A) \neq P(B)$. The probability that the family takes a summer vacation depends on whether or not they own a computer.

17. Let A = {pregnant} and B = {multiple births}. Thus $P(A) = 0.35$ and $P(B|A) = 0.28$.

    (a) $P(A \text{ and } B) = P(A) \cdot P(B|A) = (0.35) \cdot (0.28) = 0.098$

    (b) $P(B'|A) = 1 - P(B|A) = 1 - 0.28 = 0.72$

    (c) It is not unusual because the probability of a pregnancy and multiple births is 0.098.

19. Let A = {household in U.S. has a computer} and B = {has Internet access}.

$P(A \text{ and } B) = P(A)P(B|A) = (0.62)(0.88) = 0.546$

21. Let A = {1st person is left-handed} and B = {2nd person is left-handed}.

(a) $P(A \text{ and } B) = P(A) \cdot P(A|B) = \left(\dfrac{120}{1000}\right) \cdot \left(\dfrac{119}{999}\right) \approx 0.014$

(b) $P(A' \text{ and } B') = P(A') \cdot P(B'|A') = \left(\dfrac{880}{1000}\right) \cdot \left(\dfrac{879}{999}\right) \approx 0.774$

(c) $P(\text{at least one is left-handed}) = 1 - P(A' \text{ and } B') = 1 - 0.774 = 0.226$

23. Let A = {have one month's income or more} and B = {man}.

(a) $P(A) = \dfrac{138}{287} \approx 0.481$

(b) $P(A'|B) = \dfrac{66}{142} \approx 0.465$

(c) $P(B'|A) = \dfrac{62}{138} \approx 0.449$

(d) Dependent because $P(A') \approx 0.519 \neq 0.465 \approx P(A'|B)$

Whether a person has at least one month's income saved depends on whether or not the person is male.

25. (a) $P(\text{all five have AB+}) = (0.03) \cdot (0.03) \cdot (0.03) \cdot (0.03) \cdot (0.03) = 0.0000000243$

(b) $P(\text{none have AB+}) = (0.97) \cdot (0.97) \cdot (0.97) \cdot (0.97) \cdot (0.97) \approx 0.859$

(c) $P(\text{at least one has AB+}) = 1 - P(\text{none have AB+}) = 1 - 0.859 = 0.141$

27. (a) $P(\text{first question correct}) = 0.2$

(b) $P(\text{first two questions correct}) = (0.2) \cdot (0.2) = 0.04$

(c) $P(\text{first three questions correct}) = (0.2)^3 = 0.008$

(d) $P(\text{none correct}) = (0.8)^3 = 0.512$

(e) $P(\text{at least one correct}) = 1 - P(\text{none correct}) = 1 - 0.512 = 0.488$

29. $P \text{ (all three products came from the third factory)} = \dfrac{25}{110} \cdot \dfrac{24}{109} \cdot \dfrac{23}{108} \approx 0.011$

31. $P(A|B) = \dfrac{P(A) \cdot P(B|A)}{P(A) \cdot P(B|A) + P(A') \cdot P(B|A')}$

$= \dfrac{\left(\dfrac{2}{3}\right) \cdot \left(\dfrac{1}{5}\right)}{\left(\dfrac{2}{3}\right) \cdot \left(\dfrac{1}{5}\right) + \left(\dfrac{1}{3}\right) \cdot \left(\dfrac{1}{2}\right)} = \dfrac{0.133}{0.133 + 0.167} = 0.444$

33. $P(A|B) = \dfrac{P(A)P(B|A)}{(A)P(B|A) + P(A')P(B|A')}$

$= \dfrac{(0.25)(0.3)}{(0.25)(0.3) + (0.75)(0.5)} = \dfrac{0.075}{0.075 + 0.375} = 0.167$

35. $P(A) = \dfrac{1}{200} = 0.005$

$P(B|A) = 0.80$

$P(B|A') = 0.05$

(a) $P(A|B) = \dfrac{P(A) \cdot P(B|A)}{P(A) \cdot P(B|A) + P(A') \cdot P(B|A')}$

$= \dfrac{(0.005) \cdot (0.8)}{(0.005) \cdot (0.8) + (0.995) \cdot (0.05)} = \dfrac{0.004}{0.004 + 0.04975} \approx 0.074$

(b) $P(A'|B') = \dfrac{P(A') \cdot P(B'|A')}{P(A') \cdot P(B'|A') + P(A) \cdot P(B'|A)}$

$= \dfrac{(0.995) \cdot (0.95)}{(0.995) \cdot (0.95) + (0.005) \cdot (0.2)} = \dfrac{0.94525}{0.94525 + 0.001} \approx 0.999$

37. Let $A = \{\text{flight departs on time}\}$ and $B = \{\text{flight arrives on time}\}$.

$P(A|B) = \dfrac{P(A \text{ and } B)}{P(B)} = \dfrac{(0.83)}{(0.87)} \approx 0.954$

3.3 THE ADDITION RULE

3.3 Try It Yourself Solutions

1a. (1) None of the statements are true.

(2) None of the statements are true.

(3) All of the statements are true.

b. (1) A and B are not mutually exclusive.

(2) A and B are not mutually exclusive.

(3) A and B are mutually exclusive.

2a. (1) Mutually exclusive   (2) Not mutually exclusive

b. (1) Let $A = \{6\}$ and $B = \{\text{odd}\}$.

$P(A) = \dfrac{1}{6}$ and $P(B) = \dfrac{3}{6} = \dfrac{1}{2}$

(2) Let $A = \{\text{face card}\}$ and $B = \{\text{heart}\}$.

$P(A) = \dfrac{12}{52}$, $P(B) = \dfrac{13}{52}$, and $P(A \text{ and } B) = \dfrac{3}{52}$

c. (1) $P(A \text{ or } B) = P(A) + P(B) = \dfrac{1}{6} + \dfrac{1}{2} \approx 0.667$

(2) $P(A \text{ or } B) = P(A) + P(B) - P(A \text{ and } B) = \dfrac{12}{52} + \dfrac{13}{52} - \dfrac{3}{52} \approx 0.423$

3a. $A = \{$sales between \$0 and \$24,999$\}$

$B = \{$sales between \$25,000 and \$49,000$\}$

b. A and B cannot occur at the same time $\rightarrow A$ and B are mutually exclusive

c. $P(A) = \dfrac{3}{36}$ and $P(B) = \dfrac{5}{36}$

d. $P(A \text{ or } B) = P(A) + P(B) = \dfrac{3}{36} + \dfrac{5}{36} \approx 0.222$

4a. (1) Mutually exclusive   (2) Not mutually exclusive

b. (1) $P(B \text{ or } AB) = P(B) + P(AB) = \dfrac{45}{409} + \dfrac{16}{409} \approx 0.149$

(2) $P(O \text{ or } Rh+) = P(O) + P(Rh+) - P(O \text{ and } Rh+) = \dfrac{184}{409} + \dfrac{344}{409} - \dfrac{156}{409} \approx 0.910$

5a. Let $A = \{$linebacker$\}$ and $B = \{$quarterback$\}$.

$P(A \text{ or } B) = P(A) + P(B) = \dfrac{32}{255} + \dfrac{11}{255} \approx 0.169$

b. $P($not a linebacker or quarterback$) = 1 - P(A \text{ or } B) \approx 1 - 0.169 \approx 0.831$

3.3 EXERCISE SOLUTIONS

1. $P(A \text{ and } B) = 0$ because A and B cannot occur at the same time.

3. True

5. False, $P(A \text{ or } B) = P(A) + P(B) - P(A \text{ and } B)$

7. Not mutually exclusive because the two events can occur at the same time.

9. Not mutually exclusive because the two events can occur at the same time. The worker can be female and have a college degree.

11. Mutually exclusive because the two events cannot occur at the same time. The person cannot be in both age classes.

13. (a) No, it is possible for the events $\{$overtime$\}$ and $\{$temporary help$\}$ to occur at the same time.

(b) $P(\text{OT or temp}) = P(\text{OT}) + P(\text{temp}) - P(\text{OT and temp}) = \dfrac{18}{52} + \dfrac{9}{52} - \dfrac{5}{52} \approx 0.423$

15. (a) Not mutually exclusive because the two events can occur at the same time. A carton can have a puncture and a smashed corner.

(b) $P(\text{puncture or corner}) = P(\text{puncture}) + P(\text{corner}) - P(\text{puncture and corner})$

$$= 0.05 + 0.08 - 0.004 = 0.126$$

17. (a) $P(\text{diamond or } 7) = P(\text{diamond}) + P(7) \cdot P(\text{diamond and } 7)$

$$= \frac{13}{52} + \frac{4}{52} - \frac{1}{52} \approx 0.308$$

(b) $P(\text{red or queen}) = P(\text{red}) + P(\text{queen}) - P(\text{red and queen})$

$$= \frac{26}{52} + \frac{4}{52} - \frac{2}{52} \approx 0.538$$

(c) $P(3 \text{ or face card}) = P(3) + P(\text{face card}) - P(3 \text{ and face card})$

$$= \frac{4}{52} + \frac{12}{52} - 0 \approx 0.308$$

19. (a) $P(\text{under 5}) = 0.068$

(b) $P(\text{not } 65+) = 1 - P(65+) = 1 - 0.147 = 0.853$

(c) $P(\text{between 18 and 34}) = P(\text{between 18 and 24 or between 25 and 34})$
$= P(\text{between 18 and 24}) + P(\text{between 25 and 34}) = 0.097 + 0.132 = 0.229$

21. (a) $P(\text{not completely satisfied}) = 1 - \frac{102}{1017} = 0.900$

(b) $P(\text{somewhat satisfied or completely satisfied})$

$= P(\text{somewhat satisfied}) + P(\text{completely dissatisfied})$

$$= \frac{326}{1017} + \frac{132}{1017} = \frac{458}{1017} \approx 0.450$$

23. $A = \{\text{male}\}; B = \{\text{nursing major}\}$

(a) $P(A \text{ or } B) = P(A) + P(B) - P(A \text{ and } B)$

$$= \frac{1110}{3537} + \frac{795}{3537} - \frac{95}{3537} = 0.512$$

(b) $P(A' \text{ or } B') = P(A') + P(B') - P(A' \text{ and } B')$

$$= \frac{2427}{3537} + \frac{2742}{3537} - \frac{1727}{3537} = 0.973$$

(c) $P(A \text{ or } B) = 0.512$

(d) No mutually exclusive. A male can be a nursing major.

25. $A = \{\text{frequently}\}; B = \{\text{occasionally}\}; C = \{\text{not at all}\}; D = \{\text{male}\}$

(a) $P(A \text{ or } B) = \frac{428}{2850} + \frac{886}{2850} = 0.461$

(b) $P(D' \text{ or } C) = P(D') + P(C) - P(D' \text{ and } C)$

$$= \frac{1378}{2850} + \frac{1536}{2850} - \frac{741}{2850} = 0.762$$

(c) $P(D \text{ or } A) = P(D) + P(A) - P(D \text{ and } A)$

$$= \frac{1472}{2850} + \frac{428}{2850} - \frac{221}{2850} = 0.589$$

(d) $P(D' \text{ or } A') = P(D') + P(A') - P(D' \text{ and } A')$

$$= \frac{1378}{2850} + \frac{2422}{2850} - \frac{1171}{2850} = 0.922$$

(e) Not mutually exclusive. A female can be frequently involved in charity work.

27. Answers will vary.

Conclusion: If two events, $\{A\}$ and $\{B\}$, are independent, $P(A \text{ and } B) = P(A) \cdot P(B)$. If two events are mutually exclusive, $P(A \text{ and } B) = 0$. The only scenario when two events can be independent and mutually exclusive is if $P(A) = 0$ or $P(B) = 0$.

29. $P(A \text{ or } B \text{ or } C) = P(A) + P(B) + P(C) - P(A \text{ and } B) - P(A \text{ and } C) - P(B \text{ and } C)$

$$+ P(A \text{ and } B \text{ and } C)$$

$$= 0.38 + 0.26 + 0.14 - 0.12 - 0.03 - 0.09 + 0.01$$

$$= 0.55$$

3.4 COUNTING PRINCIPLES

3.4 Try It Yourself Solutions

1a. $n = 6$ teams   **b.** $6! = 720$

2a. $_8P_3 = \dfrac{8!}{(8-3)!} = \dfrac{8!}{5!} = \dfrac{8 \cdot 7 \cdot 6 \cdot 5 \cdot 4 \cdot 3 \cdot 2 \cdot 1}{5 \cdot 4 \cdot 3 \cdot 2 \cdot 1} = 8 \cdot 7 \cdot 6 = 336$

b. There are 336 possible ways that the subject can pick a first, second, and third activity.

3a. $n = 12, r = 4$

b. $_{12}P_4 = \dfrac{12!}{(12-4)!} = \dfrac{12!}{8!} = 12 \cdot 11 \cdot 10 \cdot 9 = 11{,}880$

4a. $n = 20, n_1 = 6, n_2 = 9, n_3 = 5$

b. $\dfrac{n!}{n_1! \cdot n_2! \cdot n_3!} = \dfrac{20!}{6! \cdot 9! \cdot 5!} = 77{,}597{,}520$

5a. $n = 20, r = 3$

b. $_{20}C_3 = 1140$

c. There are 1140 different possible three-person committees that can be selected from 20 employees.

6a. $_{20}P_2 = 380$ **b.** $\dfrac{1}{380} \approx 0.003$

7a. 1 favorable outcome and $\dfrac{6!}{1! \cdot 2! \cdot 2! \cdot 1!} = 180$ distinguishable permutations.

b. $P(\text{Letter}) = \dfrac{1}{180} \approx 0.006$

8a. $_{15}C_5 = 3003$ **b.** $_{54}C_5 = 3,162,510$ **c.** 0.0009

9a. $(_5C_3) \cdot (_7C_0) = 10 \cdot 1 = 10$ **b.** $_{12}C_3 = 220$ **c.** $\dfrac{10}{220} \approx 0.045$

3.4 EXERCISE SOLUTIONS

1. The number of ordered arrangements of n objects taken r at a time. An exmple of a permutation is the number of seating arrangements of you and three friends.

3. False, a permutation is an ordered arrangement of objects.

5. True

7. $_7P_3 = \dfrac{7!}{(7-3)!} = \dfrac{7!}{4!} = \dfrac{5040}{24} = 210$

9. $_7C_4 = \dfrac{7!}{4!(7-4)!} = \dfrac{(7)(6)(5)(4)(3)(2)(1)}{[(4)(3)(2)(1)][(3)(2)(1)]} = \dfrac{540}{(24)(6)} = 35$

11. $_{24}C_6 = \dfrac{24!}{6!(24-6)!} = \dfrac{24!}{6!18!} = 134,596$

13. $\dfrac{_6P_2}{_{10}P_4} = \dfrac{\dfrac{6!}{(6-2)!}}{\dfrac{10!}{(10-4)!}} = \dfrac{\dfrac{6!}{4!}}{\dfrac{10!}{6!}} = \dfrac{\dfrac{720}{24}}{\dfrac{3,628,800}{720}} = 0.0060$

15. Permutation, because order of the 15 people in line matters.

17. Combinations, because the order of the captains does not matter.

19. $10 \cdot 8 \cdot _{13}C_2 = 6240$

21. $6! = 720$

23. $_{52}C_6 = 20,358,520$

25. $\dfrac{18!}{4! \cdot 8! \cdot 6!} = 9,189,180$

27. 3-S's, 3-T's, 1-A, 2-I's, 1-C

$\dfrac{10!}{3! \cdot 3! \cdot 1! \cdot 2! \cdot 1!} = 50,400$

29. (a) $6! = 720$

(b) sample

(c) $\dfrac{1}{720} = 0.0014$

31. (a) 12

(b) tree

(c) $\dfrac{1}{12} = 0.0833$

33. (a) 907,200

(b) population

(c) $\dfrac{1}{907,200} = 0.000001$

35. $\dfrac{1}{_{12}C_3} = \dfrac{1}{220} = 0.0045$

37. (a) $\left(\dfrac{15}{56}\right)\left(\dfrac{14}{55}\right)\left(\dfrac{13}{54}\right) = 0.0164$

    (b) $\left(\dfrac{41}{56}\right)\left(\dfrac{40}{55}\right)\left(\dfrac{39}{54}\right) = 0.385$

39. (a) $_8C_4 = 70$

    (b) $2 \cdot 2 \cdot 2 \cdot 2 = 16$

    (c) $_4C_2\left[\dfrac{(_2C_0) \cdot (_2C_0) \cdot (_2C_2) \cdot (_2C_2)}{_8C_4}\right] \approx 0.086$

41. (a) $(26)(26)(10)(10)(10)(10)(10) = 67{,}600{,}000$

    (b) $(26)(25)(10)(9)(8)(7)(6) = 19{,}656{,}000$

    (c) $\dfrac{1}{67{,}600{,}000} \approx 0.000000015$

43. (a) $5! = 120$ (b) $2! \cdot 3! = 12$ (c) $3! \cdot 2! = 12$ (d) 0.4

45. $(8\%)(1200) = (0.08)(1200) = 96$ of the 1200 rate financial shape as excellent.

$P(\text{all four rate excellent}) = \dfrac{_{96}C_4}{_{1200}C_4} = \dfrac{3{,}321{,}960}{85{,}968{,}659{,}700} \approx 0.000039$

47. $(36\%)(500) = (0.36)(500) = 180$ of the 1500 rate financial shape as fair $\Rightarrow 500 - 180 = 320$ rate shape as not fair.

$P(\text{none of 80 selected rate fair}) = \dfrac{_{320}C_{80}}{_{500}C_{80}} \approx 5.03 \times 10^{-18}$

49. (a) $_{40}C_5 = 658{,}008$

    (b) $P(\text{win}) = \dfrac{1}{658{,}008} \approx 0.00000152$

51. (a) $\dfrac{_{13}C_1 \, _4C_4 \, _{12}C_1 \, _4C_1}{_{52}C_5} = \dfrac{(13)(1)(12)(4)}{2{,}598{,}960} = 0.0002$

    (b) $\dfrac{_{13}C_1 \, _4C_3 \, _{12}C_1 \, _4C_2}{_{52}C_5} = \dfrac{(13)(4)(12)(6)}{2{,}598{,}960} = 0.00144$

    (c) $\dfrac{_{13}C_1 \, _4C_3 \, _{12}C_2 \, _4C_1 \, _4C_1}{_{52}C_5} = \dfrac{(13)(4)(66)(4)(4)}{2{,}598{,}960} = 0.0211$

    (d) $\dfrac{_{13}C_2 \, _{13}C_1 \, _{13}C_1 \, _{13}C_1}{_{52}C_5} = \dfrac{(78)(13)(13)(13)}{2{,}598{,}960} = 0.0659$

53. $_{14}C_4 = 1001$ possible 4 digit arrangements if order is not important.

Assign 1000 of the 4 digit arrangements to the 13 teams since 1 arrangement is excluded.

55. $P(\text{1st}) = \dfrac{250}{1000} = 0.250$ $\qquad$ $P(\text{8th}) = \dfrac{28}{1000} = 0.028$

$\qquad$ $P(\text{2nd}) = \dfrac{199}{1000} = 0.199$ $\qquad$ $P(\text{9th}) = \dfrac{17}{1000} = 0.017$

$\qquad$ $P(\text{3rd}) = \dfrac{156}{1000} = 0.156$ $\qquad$ $P(\text{10th}) = \dfrac{11}{1000} = 0.011$

$\qquad$ $P(\text{4th}) = \dfrac{119}{1000} = 0.119$ $\qquad$ $P(\text{11th}) = \dfrac{8}{1000} = 0.008$

$\qquad$ $P(\text{5th}) = \dfrac{88}{1000} = 0.088$ $\qquad$ $P(\text{12th}) = \dfrac{7}{1000} = 0.007$

$\qquad$ $P(\text{6th}) = \dfrac{63}{1000} = 0.063$ $\qquad$ $P(\text{13th}) = \dfrac{6}{1000} = 0.006$

$\qquad$ $P(\text{7th}) = \dfrac{43}{1000} = 0.043$ $\qquad$ $P(\text{14th}) = \dfrac{5}{1000} = 0.005$

57. Let $A = \{\text{team with the worst record wins third pick}\}$ and

$\qquad$ $B = \{\text{team with the best record, ranked 13th, wins first pick}\}$ and

$\qquad$ $C = \{\text{team ranked 2nd wins the second pick}\}$.

$\qquad$ $P(A\,|\,B \text{ and } C) = \dfrac{250}{796} = 0.314$

CHAPTER 3 REVIEW EXERCISE SOLUTIONS

1. Sample space:

$\qquad$ {HHHH, HHHT, HHTH, HHTT, HTHH, HTHT, HTTH, HTTT, THHH, THHT, THTH, THTT, TTHH, TTHT, TTTH, TTTT}

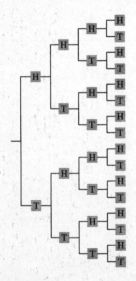

$\qquad$ Event: Getting three heads

$\qquad$ {HHHT, HHTH, HTHH, THHH}

3. Sample space: {Jan, Feb . . ., Dec}

Event: {Jan, June, July}

5. $(7)(4)(3) = 84$

7. Empirical probability

9. Subjective probability

11. Classical probability

13. $P(\text{at least } 10) = 0.108 + 0.089 + 0.018 = 0.215$

15. $\dfrac{1}{(8)(10)(10)(10)(10)(10)(10)} = 1.25 \times 10^{-7}$

17. $P(\text{undergrad} \mid +) = 0.92$

19. Independent, the first event does not affect the outcome of the second event.

21. $P(\text{correct toothpaste and correct dental rinse}) = P(\text{correct toothpaste}) \cdot P(\text{correct dental rinse})$

$$= \left(\frac{1}{8}\right) \cdot \left(\frac{1}{5}\right) \approx 0.025$$

23. Mutually exclusive because both events cannot occur at the same time.

25. $P(\text{home or work}) = P(\text{home}) + P(\text{work}) - P(\text{home and work}) = 0.44 + 0.37 - 0.21 = 0.60$

27. $P(4\text{–}8 \text{ or club}) = P(4\text{–}8) + P(\text{club}) - P(4\text{–}8 \text{ and club}) = \dfrac{20}{52} + \dfrac{13}{52} - \dfrac{5}{52} \approx 0.538$

29. $P(\text{odd or less than } 4) = P(\text{odd}) + P(\text{less than } 4) - P(\text{odd and less than } 4)$

$$= \frac{6}{12} + \frac{3}{12} - \frac{2}{12} \approx 0.583$$

31. $P(600 \text{ or more}) = P(600 - 999) + P(1000 \text{ or more})$

$$= 0.198 + 0.300 = 0.498$$

33. $P(\text{poor taste or hard to find}) = P(\text{poor taste}) + P(\text{hard to find})$

$$= \frac{60}{500} + \frac{55}{500} = \frac{115}{500} = 0.23$$

35. Order is important: $_{15}P_3 = 2730$

37. Order is not important: $_{17}C_4 = 2380$

39. $P(3 \text{ kings and 2 queens}) = \dfrac{_4C_3 \cdot \, _4C_2}{_{52}C_5} = \dfrac{4 \cdot 6}{2{,}598{,}960} \approx 0.00000923$

41. (a) $P(\text{no defectives}) = \dfrac{_{197}C_3}{_{200}C_3} = \dfrac{1{,}254{,}890}{1{,}313{,}400} \approx 0.955$

   (b) $P(\text{all defective}) = \dfrac{_3C_3}{_{200}C_3} = \dfrac{1}{1{,}313{,}400} = \, \approx 0.000000761$

   (c) $P(\text{at least one defective}) = 1 - P(\text{no defective}) = 1 - 0.955 = 0.045$

   (d) $P(\text{at least one non-defective}) = 1 - P(\text{all defective}) = 1 - 0.000000761 \approx 0.999999239$

CHAPTER 3 QUIZ SOLUTIONS

1. (a) $P(\text{bachelor}) = \dfrac{1399}{2671} \approx 0.524$

(b) $P(\text{bachelor}|\text{F}) = \dfrac{804}{1561} \approx 0.515$

(c) $P(\text{bachelor}|\text{M}) = \dfrac{595}{1110} \approx 0.536$

(d) $P(\text{associate or bachelor}) = P(\text{associate}) + P(\text{bachelor})$

$$= \dfrac{665}{2671} + \dfrac{1399}{2671} \approx 0.773$$

(e) $P(\text{doctorate}|\text{M}) = \dfrac{25}{1110} \approx 0.023$

(f) $P(\text{master or female}) = P(\text{master}) + P(\text{female}) - P(\text{master and female})$

$$= \dfrac{559}{2671} + \dfrac{1561}{2671} - \dfrac{329}{2671} = 0.671$$

(g) $P(\text{associate and male}) = P(\text{associate}) \cdot P(\text{male}|\text{associate})$

$$= \dfrac{665}{2671} \cdot \dfrac{260}{665} \approx 0.097$$

(h) $P(\text{F}|\text{bachelor}) = \dfrac{804}{1399} \approx 0.575$

2. Not mutually exclusive because both events can occur at the same time.

Dependent because one event can affect the occurrence of the second event.

3. (a) $_{147}C_3 = 518{,}665$

(b) $_3C_3 = 1$

(c) $_{150}C_3 - {_3}C_3 = 551{,}300 - 1 = 551{,}299$

4. (a) $\dfrac{_{147}C_3}{_{150}C_3} = \dfrac{518{,}665}{551{,}300} \approx 0.94$

(b) $\dfrac{_3C_3}{_{150}C_3} = \dfrac{1}{551{,}300} \approx 0.00000181$

(c) $\dfrac{_{150}C_3 - {_3}C_3}{_{150}C_3} = \dfrac{551{,}299}{551{,}300} \approx 0.999998$

5. $9 \cdot 10 \cdot 10 \cdot 10 \cdot 10 \cdot 5 = 450{,}000$

6. $_{30}P_4 = 657{,}720$

Discrete Probability Distributions

4.1 Try It Yourself Solutions

1a. (1) measured   (2) counted

b. (1) Random variable is continuous because x can be any amount of time needed to complete a test.

(2) Random variable is discrete because x can be counted.

2ab.

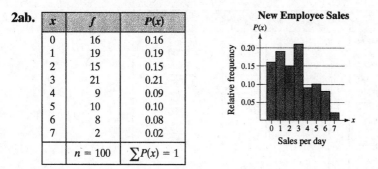

x	f	$P(x)$
0	16	0.16
1	19	0.19
2	15	0.15
3	21	0.21
4	9	0.09
5	10	0.10
6	8	0.08
7	2	0.02
	$n = 100$	$\sum P(x) = 1$

New Employee Sales

3a. Each $P(x)$ is between 0 and 1.

b. $\Sigma P(x) = 1$

c. Because both conditions are met, the distribution is a probability distribution.

4a. (1) Yes, each outcome is between 0 and 1.   (2) Yes, each outcome is between 0 and 1.

b. (1) Yes, $\Sigma P(x) = 1$.   (2) Yes, $\Sigma P(x) = 1$.

c. (1) Is a probability distribution   (2) Is a probability distribution

5ab.

x	$P(x)$	$xP(x)$
0	0.16	$(0)(0.16) = 0.00$
1	0.19	$(1)(0.19) = 0.19$
2	0.15	$(2)(0.15) = 0.30$
3	0.21	$(3)(0.21) = 0.63$
4	0.09	$(4)(0.09) = 0.36$
5	0.10	$(5)(0.10) = 0.50$
6	0.08	$(6)(0.08) = 0.48$
7	0.02	$(7)(0.02) = 0.14$
	$\sum P(x) = 1$	$\sum xP(x) = 2.60$

c. $\mu = \Sigma xP(x) = 2.6$

On average, 2.6 sales are made per day.

6ab.

x	$P(x)$	$x - \mu$	$(x - \mu)^2$	$P(x)(x - \mu)^2$
0	0.16	−2.6	6.76	(0.16)(6.76) = 1.0816
1	0.19	−1.6	2.56	(0.19)(2.56) = 0.4864
2	0.15	−0.6	0.36	(0.15)(0.36) = 0.054
3	0.21	0.4	0.16	(0.21)(0.16) = 0.0336
4	0.09	1.4	1.96	(0.09)(1.96) = 0.1764
5	0.10	2.4	5.76	(0.10)(5.76) = 0.576
6	0.08	3.4	11.56	(0.08)(11.56) = 0.9248
7	0.02	4.4	19.36	(0.02)(19.36) = 0.3872
	$\Sigma P(x) = 1$			$\Sigma P(x)(x - \mu)^2 = 3.72$

c. $\sigma = \sqrt{\sigma^2} = \sqrt{3.720} \approx 1.9$

d. A typical distance or deviation of the random variable from the mean is 1.9 sales per day.

7ab.

x	f	$P(x)$	$xP(x)$
0	25	0.111	(0)(0.111) = 0.000
1	48	0.213	(1)(0.213) = 0.213
2	60	0.267	(2)(0.267) = 0.533
3	45	0.200	(3)(0.200) = 0.600
4	20	0.089	(4)(0.089) = 0.356
5	10	0.044	(5)(0.044) = 0.222
6	8	0.036	(6)(0.036) = 0.213
7	5	0.022	(7)(0.022) = 0.156
8	3	0.013	(8)(0.013) = 0.107
9	1	0.004	(9)(0.004) = 0.040
	$n = 225$	$\Sigma P(x) \approx 1$	$\Sigma xP(x) = 2.440$

Gain, x	$P(x)$	$xP(x)$
1995	$\dfrac{1}{2000}$	$\dfrac{1995}{2000}$
995	$\dfrac{1}{2000}$	$\dfrac{995}{2000}$
495	$\dfrac{1}{2000}$	$\dfrac{495}{2000}$
245	$\dfrac{1}{2000}$	$\dfrac{245}{2000}$
95	$\dfrac{1}{2000}$	$\dfrac{95}{2000}$
−5	$\dfrac{1995}{2000}$	$-\dfrac{9975}{2000}$
	$\Sigma P(x) = 1$	$\Sigma xP(x) = -3.08$

c. $E(x) = \Sigma xP(x) = -3.08$

d. You can expect an average loss of $3.08 for each ticket purchased.

4.1 EXERCISE SOLUTIONS

1. A random variable represents a numerical value assigned to an outcome of a probability experiment. Examples: Answers will vary.

3. An expected value of 0 represents the break even point, so the accountant will not gain or lose any money.

5. False. In most applications, discrete random variables represent counted data, while continuous random variables represent measured data.

7. True

9. Discrete, because home attendance is a random variable that is countable.

11. Continuous, because annual vehicle-miles driven is a random variable that cannot be counted.

13. Discrete, because the number of motorcycle accidents is a random variable that is countable.

15. Continuous, because the random variable has an infinite number of possible outcomes and cannot be counted.

17. Discrete, because the random variable is countable.

19. Continuous, because the random variable has an infinite number of possible outcomes and cannot be counted.

21. (a) $P(x > 2) = 0.25 + 0.10 = 0.35$

   (b) $P(x < 4) = 1 - P(4) = 1 - 0.10 = 0.90$

23. $\Sigma P(x) = 1 \rightarrow P(3) = 0.22$

25. Yes

27. No, $\Sigma P(x) = 0.95$ and $P(5) < 0$.

29. (a)

x	f	$P(x)$	$xP(x)$	$(x - \mu)$	$(x - \mu)^2$	$(x - \mu)^2 P(x)$
0	1491	0.686	$(0)(0.686) = 0$	-0.501	0.251	$(0.251)(0.686) = 0.169$
1	425	0.195	$(1)(0.195) = 0.195$	0.499	0.249	$(0.249)(0.195) = 0.049$
2	168	0.077	$(2)(0.077) = 0.155$	1.499	2.246	$(2.246)(0.077) = 0.173$
3	48	0.022	$(3)(0.022) = 0.066$	2.499	6.244	$(6.244)(0.022) = 0.138$
4	29	0.013	$(4)(0.013) = 0.053$	3.499	12.241	$(12.241)(0.013) = 0.160$
5	14	0.006	$(5)(0.006) = 0.030$	4.499	20.238	$(20.238)(0.006) = 0.122$
	$n = 2175$	$\sum P(x) \approx 1$	$\sum xP(x) = 0.497$			$\sum (x - \mu)^2 P(x) = 0.811$

   (b) $\mu = \Sigma xP(x) \approx 0.5$

   (c) $\sigma^2 = \Sigma(x - \mu)^2 P(x) \approx 0.8$

   (d) $\sigma = \sqrt{\sigma^2} \approx \sqrt{0.8246} \approx 0.9$

   (e) A household on average has 0.5 dog with a standard deviation of 0.9 dog.

31. (a)

x	f	$P(x)$	$xP(x)$	$(x - \mu)$	$(x - \mu)^2$	$(x - \mu)^2 P(x)$
0	300	0.432	0.000	-0.764	0.584	$(0.584)(0.432) = 0.252$
1	280	0.403	0.403	0.236	0.056	$(0.056)(0.403) = 0.022$
2	95	0.137	0.274	1.236	1.528	$(1.528)(0.137) = 0.209$
3	20	0.029	0.087	2.236	5.000	$(5.000)(0.029) = 0.145$
	$n = 695$	$\sum P(x) \approx 1$	$\sum xP(x) = 0.764$			$\sum (x - \mu)^2 P(x) = 0.629$

   (b) $\mu = \Sigma xP(x) \approx 0.8$

   (c) $\sigma^2 = \Sigma(x - \mu)^2 P(x) \approx 0.6$

   (d) $\sigma = \sqrt{\sigma^2} \approx 0.8$

   (e) A household on average has 0.8 computer with a standard deviation of 0.8 computer.

33. (a)

x	f	$P(x)$	$xP(x)$	$(x - \mu)$	$(x - \mu)^2$	$(x - \mu)^2 P(x)$
0	6	0.031	0.000	-3.411	11.638	$(11.638)(0.031) = 0.360$
1	12	0.063	0.063	-2.411	5.815	$(5.815)(0.063) = 0.366$
2	29	0.151	0.302	-1.411	1.992	$(1.992)(0.151) = 0.300$
3	57	0.297	0.891	-0.411	0.169	$(0.169)(0.297) = 0.050$
4	42	0.219	0.876	0.589	0.346	$(0.346)(0.219) = 0.076$
5	30	0.156	0.780	1.589	2.523	$(2.523)(0.156) = 0.394$
6	16	0.083	0.498	2.589	6.701	$(6.701)(0.083) = 0.557$
	$n = 192$	$\sum P(x) = 1$	$\sum xP(x) = 3.410$			$\sum (x - \mu)^2 P(x) = 2.103$

   (b) $\mu = \Sigma xP(x) \approx 3.4$

   (c) $\sigma^2 = \Sigma(x - \mu)^2 P(x) \approx 2.1$

(d) $\sigma = \sqrt{\sigma^2} \approx 1.5$

(e) An employee works an average of 3.4 overtime hours per week with a standard deviation of 1.5 hours.

35.

x	$P(x)$	$xP(x)$	$(x - \mu)$	$(x - \mu)^2$	$(x - \mu)^2 P(x)$
0	0.02	0.00	−5.30	28.09	0.562
1	0.02	0.02	−4.30	18.49	0.372
2	0.06	0.12	−3.30	10.89	0.653
3	0.06	0.18	−2.30	5.29	0.317
4	0.08	0.32	−1.30	1.69	0.135
5	0.22	1.10	−0.30	0.09	0.020
6	0.30	1.80	0.70	0.49	0.147
7	0.16	1.12	1.70	2.89	0.462
8	0.08	0.64	2.70	7.29	0.583
	$\sum P(x) = 1$	$\sum xP(x) = 5.30$			$\sum (x - \mu)^2 P(x) = 3.250$

(a) $\mu = \Sigma x P(x) = 5.3$

(b) $\sigma^2 = \Sigma (x - \mu)^2 P(x) = 3.3$

(c) $\sigma = \sqrt{\sigma^2} = 1.8$

(d) $E[x] = \mu = \Sigma x P(x) = 5.3$

(e) The expected number of correctly answered questions is 5.3 with a standard deviation of 1.8 questions.

37. (a) $\mu = \Sigma x P(x) \approx 2.0$ (b) $\sigma^2 = \Sigma (x - \mu)^2 P(x) \approx 1.0$

(c) $\sigma = \sqrt{\sigma^2} \approx 1.0$ (d) $E[x] = \mu = \Sigma x P(x) = 2.0$

(e) The expected number of hurricanes that hit the U.S. is 2.0 with a standard deviation of 1.0.

39. (a) $\mu = \Sigma x P(x) \approx 2.5$ (b) $\sigma^2 = \Sigma (x - \mu)^2 P(x) \approx 1.9$

(c) $\sigma = \sqrt{\sigma^2} \approx 1.4$ (d) $E[x] = \mu = \Sigma x P(x) = 2.5$

(e) The expected household size is 2.5 persons with a standard deviation of 1.4 persons.

41. (a) $P(x < 2) = 0.686 + 0.195 = 0.881$

(b) $P(x \geq 1) = 1 - P(x = 0) = 1 - 0.686 = 0.314$

(c) $P(1 \leq x \leq 3) = 0.195 + 0.077 + 0.022 = 0.294$

43. A household with three dogs is unusual because the probability is only 0.022.

45. $E(x) = \mu = \Sigma x P(x) = (-1) \cdot \left(\frac{37}{38}\right) + (35) \cdot \left(\frac{1}{38}\right) \approx -\$0.05$

47. $\mu_4 = a + b\mu_x = 1000 + 1.05(36{,}000) = \$38{,}800$

49. $\mu_{x+y} = \mu_x + \mu_y = 1532 + 1506 = 3038$

$\mu_{x-y} = \mu_x - \mu_y = 1532 - 1506 = 26$

4.2 Try It Yourself Solutions

1a. Trial: answering a question (10 trials)
Success: question answered correctly

b. Yes, the experiment satisfies the four conditions of a binomial experiment.

c. It is a binomial experiment.

$n = 10, p = 0.25, q = 0.75, x = 0, 1, 2, \ldots, 9, 10$

2a. Trial: drawing a card with replacement (5 trials)
Success: card drawn is a club
Failure: card drawn is not a club

b. $n = 5, p = 0.25, q = 0.75, x = 3$

c. $P(3) = \dfrac{5!}{2!3!}(0.25)^3(0.75)^2 \approx 0.088$

3a. Trial: selecting a worker and asking a question (7 trials)
Success: Selecting a worker who will rely on pension
Failure: Selecting a worker who will not rely on pension

b. $n = 7, p = 0.26, q = 0.74, x = 0, 1, 2, \ldots, 6, 7$

c. $P(0) = {}_7C_0(0.26)^0(0.74)^7 = 0.1215$
$P(1) = {}_7C_1(0.26)^1(0.74)^6 = 0.2989$
$P(2) = {}_7C_2(0.26)^2(0.74)^5 = 0.3150$
$P(3) = {}_7C_3(0.26)^3(0.74)^4 = 0.1845$
$P(4) = {}_7C_4(0.26)^4(0.74)^3 = 0.0648$
$P(5) = {}_7C_5(0.26)^5(0.74)^2 = 0.0137$
$P(6) = {}_7C_6(0.26)^6(0.74)^1 = 0.0016$
$P(7) = {}_7C_7(0.26)^7(0.74)^0 = 0.0001$

d.

x	$P(x)$
0	0.1215
1	0.2989
2	0.3150
3	0.1845
4	0.0648
5	0.0137
6	0.0016
7	0.0001

4a. $n = 250, p = 0.71, x = 178$

b. $P(178) \approx 0.056$

c. The probability that exactly 178 people from a random sample of 250 people in the United States will use more than one topping on their hot dog is about 0.056.

5a. (1) $x = 2$ (2) $x = 2, 3, 4,$ or 5 (3) $x = 0$ or 1

b. (1) $P(2) \approx 0.217$

(2) $P(0) = {}_5C_0(0.21)^0(0.79)^5 = 0.308$

$P(1) = {}_5C_1(0.21)^1(0.79)^4 = 0.409$

$P(x \geq 2) = 1 - P(0) - P(1) = 1 - 0.308 - 0.409 = 0.283$

or

$P(x \geq 2) = P(2) + P(3) + P(4) + P(5)$

$= 0.217 + 0.058 + 0.008 + 0.0004$

$= 0.283$

(3) $P(x < 2) = P(0) + P(1) = 0.308 + 0.409 = 0.717$

c. (1) The probability that exactly two of the five men consider fishing their favorite leisure-time activity is about 0.217.

(2) The probability that at least two of the five men consider fishing their favorite leisure-time activity is about 0.283.

(3) The probability that fewer than two of the five men consider fishing their favorite leisure-time activity is about 0.717.

6a. Trial: selecting a business and asking if it has a Web site. (10 trials)

Success: Selecting a business with a Web site

Failure: Selecting a business without a site

b. $n = 10, p = 0.45, x = 4$

c. $P(4) \approx 0.238$

d. The probability of randomly selecting 10 small businesses and finding exactly 4 that have a website is 0.238.

7a. $P(0) = {}_6C_0(0.62)^0(0.38)^6 = 1(0.62)^0(0.38)^6 = 0.003$

$P(1) = {}_6C_1(0.62)^1(0.38)^5 = 6(0.62)^1(0.38)^5 = 0.029$

$P(2) = {}_6C_2(0.62)^0(0.38)^4 = 15(0.62)^2(0.38)^4 = 0.120$

$P(3) = {}_6C_3(0.62)^3(0.38)^3 = 20(0.62)^3(0.38)^3 = 0.262$

$P(4) = {}_6C_4(0.62)^4(0.38)^2 = 15(0.62)^4(0.38)^2 = 0.320$

$P(5) = {}_6C_5(0.62)^5(0.38)^1 = 6(0.62)^5(0.38)^1 = 0.209$

$P(6) = {}_6C_6(0.62)^6(0.38)^0 = 1(0.62)^6(0.38)^0 = 0.057$

b.

x	$P(x)$
0	0.003
1	0.029
2	0.120
3	0.262
4	0.320
5	0.209
6	0.057

c.

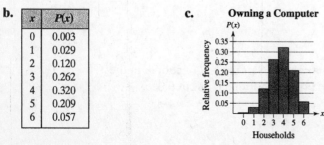

Owning a Computer

8a. Success: Selecting a clear day

$n = 31, p = 0.44, q = 0.56$

b. $\mu = np = (31)(0.44) = 13.6$

c. $\sigma^2 = npq = (31)(0.44)(0.56) = 7.6$

d. $\sigma = \sqrt{\sigma^2} = 2.8$

e. On average, there are about 14 clear days during the month of May. The standard deviation is about 3 days.

4.2 EXERCISE SOLUTIONS

1. (a) $p = 0.50$ (graph is symmetric)

(b) $p = 0.20$ (graph is skewed right $\rightarrow p < 0.5$)

(c) $p = 0.80$ (graph is skewed left $\rightarrow p > 0.5$)

3. (a) $n = 12$, $(x = 0, 1, 2, \ldots, 12)$

(b) $n = 4$, $(x = 0, 1, 2, 3, 4)$

(c) $n = 8$, $(x = 0, 1, 2, \ldots, 8)$

As n increases, the probability distribution becomes more symmetric.

5. (a) $0, 1$ (b) $0, 5$ (c) $4, 5$

7. Is a binomial experiment.

Success: baby recovers

$n = 5, p = 0.80, q = 0.20, x = 0, 1, 2, \ldots, 5$

9. Is a binomial experiment.

Success: Person said taxcuts hurt the economy.

$n = 15, p = 0.21, q = 0.79; x = 0, 1, 2, \ldots, 15$

11. $\mu = np = (80)(0.3) = 24$

$\sigma^2 = npq = (80)(0.3)(0.7) = 16.8$

$\sigma = \sqrt{\sigma^2} = 4.1$

13. $\mu = np = (124)(0.26) = 32.24$

$\sigma^2 = npq = (124)(0.26)(0.74) = 23.858$

$\sigma = \sqrt{\sigma^2} = 4.884$

15. $n = 5, p = .25$

(a) $P(3) \approx 0.088$

(b) $P(x \geq 3) = P(3) + P(4) + P(5) = 0.088 + 0.015 + .001 = 0.104$

(c) $P(x < 3) = 1 - P(x \geq 3) = 1 - 0.104 = 0.896$

17. $n = 10, p = 0.59$ (using binomial formula)

   (a) $P(8) = 0.111$

   (b) $P(x \geq 8) = P(8) + P(9) + P(10) = 0.111 + 0.036 + 0.005 = 0.152$

   (c) $P(x < 8) = 1 - P(x \geq 8) = 1 - 0.152 = 0.848$

19. $n = 10, p = 0.21$ (using binomial formula)

   (a) $P(3) \approx 0.213$

   (b) $P(x > 3) = 1 - P(0) - P(1) - P(2) - P(3)$

   $\qquad = 1 - 0.095 - 0.252 - 3.01 - 0.213 = 0.139$

   (c) $P(x \leq 3) = 1 - P(x > 3) = 1 - 0.139 = 0.861$

21. (a) $P(3) = 0.028$

   (b) $P(x \geq 4) = 1 - P(x \leq 3)$

   $\qquad = 1 - (P(0) + P(1) + P(2) + P(3))$

   $\qquad = 1 - (0.000 + 0.001 + 0.007 + 0.028)$

   $\qquad = 0.964$

   (c) $P(x \leq 2) = P(0) + P(1) + P(2) = 0.000 + 0.001 + 0.007 = 0.008$

23. (a) $P(2) = 0.255$

   (b) $P(x > 2) = 1 - P(x \leq 2)1 - (P(0) + P(1) + P(2))$

   $\qquad = 1 - (0.037 + 0.146 + 0.255)$

   $\qquad = 0.562$

   (c) $P(2 \leq x \leq 5) = P(2) + P(3) + P(4) + P(5)$

   $\qquad = 0.255 + 0.264 + 0.180 + 0.084$

   $\qquad = 0.783$

25. (a) $n = 6, p = 0.37$ (b) **Women Baseball Fans**   (c) Skewed right

x	P(x)
0	0.063
1	0.220
2	0.323
3	0.253
4	0.112
5	0.026
6	0.003

   (d) $\mu = np = (6)(0.37) = 2.2$

   (e) $\sigma^2 = npq = (6)(0.37)(0.63) \approx 1.4$

   (f) $\sigma = \sqrt{\sigma^2} \approx 1.2$

   (g) On average, 2.2 out of 6 women would consider themselves basketball fans. The standard deviation is 1.2 women.

   $x = 0, 5,$ or 6 would be unusual due to their low probabilities.

27. (a) $n = 4, p = 0.05$ (b)   (c) Skewed right

x	$P(x)$
0	0.814506
1	0.171475
2	0.013537
3	0.000475
4	0.000006

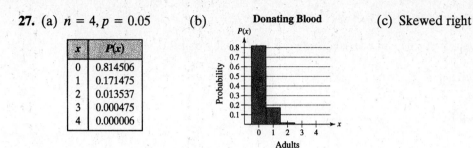

(d) $\mu = np = (4)(0.05) = 0.2$

(e) $\sigma^2 = npq = (4)(0.05)(0.95) = 0.2$

(f) $\sigma = \sqrt{\sigma^2} \approx 0.4$

(g) On average, 0.2 eligible adult out of every 4 give blood. The standard deviation is 0.4 adult.

  $x = 2, 3,$ or 4 would be uncommon due to their low probabilities.

29. (a) $n = 6, p = 0.29$ (b) $P(2) = 0.321$

x	$P(x)$
0	0.128
1	0.314
2	0.321
3	0.175
4	0.053
5	0.009
6	0.001

(c) $P(\text{at least } 5) = P(5) + P(6) = 0.009 + 0.001 = 0.010$

31. $\mu = np = (6)(0.29) = 1.7$

  $\sigma^2 = npq = (6)(0.29)(0.71) = 1.2$

  $\sigma = \sqrt{\sigma^2} = 1.1$

If 6 drivers are randomly selected, on average 1.7 drivers will name talking on cell phones as the most annoying habit of other drivers.

Five of the six or all of the six randomly selected drivers naming talking on cell phones as the most annoying habit of other drivers is rare because their probabilities are less than 0.05.

33. $P(5, 2, 2, 1) = \dfrac{10!}{5!2!2!1!}\left(\dfrac{9}{16}\right)^5\left(\dfrac{3}{16}\right)^2\left(\dfrac{3}{16}\right)^2\left(\dfrac{1}{16}\right)^1 \approx 0.033$

4.3 MORE DISCRETE PROBABILITY DISTRIBUTIONS

4.3 Try It Yourself Solutions

1a. $P(1) = (0.23)(0.77)^0 = 0.23$

  $P(2) = (0.23)(0.77)^1 = 0.177$

  $P(3) = (0.23)(0.77)^2 = 0.136$

b. $P(x < 4) = P(1) + P(2) + P(3) = 0.543$

c. The probability that your first sale will occur before your fourth sales call is 0.543.

2a. $P(0) = \dfrac{3^0(2.71828)^{-3}}{0!} \approx 0.050$

$P(1) = \dfrac{3^1(2.71828)^{-3}}{1!} \approx 0.149$

$P(2) = \dfrac{3^2(2.71828)^{-3}}{2!} \approx 0.224$

$P(3) = \dfrac{3^3(2.71828)^{-3}}{3!} \approx 0.224$

$P(4) = \dfrac{3^4(2.71828)^{-3}}{4!} \approx 0.168$

b. $P(0) + P(1) + P(2) + P(3) + P(4) \approx 0.050 + 0.149 + 0.224 + 0.224 + 0.168 \approx 0.815$

c. $1 - 0.815 \approx 0.185$

d. The probability that more than four accidents will occur in any given month at the intersection is 0.185.

3a. $\mu = \dfrac{2000}{20,000} = 0.10$

b. $\mu = 0.10, \ x = 3$

c. $P(3) = 0.0002$

d. The probability of finding three brown trout in any given cubic meter of the lake is 0.0002.

4.3 EXERCISE SOLUTIONS

1. $P(2) = (0.60)(0.4)^1 = 0.24$

3. $P(6) = (0.09)(0.91)^5 = 0.056$

5. $P(3) = \dfrac{(4)^3(e^{-4})}{3!} = 0.195$

7. $P(2) = \dfrac{(1.5)^2(e^{-1.5})}{2!} = 0.251$

9. The binomial distribution counts the number of successes in n trials. The geometric counts the number of trials until the first success is obtained.

11. Geometric. You are interested in counting the number of trials until the first success.

13. Poisson. You are interested in counting the number of occurrences that take place within a given unit of space.

15. Binomial. You are interested in counting the number of successes out of n trials.

17. $p = 0.19$

(a) $P(5) = (0.19)(0.81)^4 \approx 0.082$

(b) $P(\text{sale on 1st, 2nd, or 3rd call})$

$= P(1) + P(2) + P(3) = (0.19)(0.81)^0 + (0.19)(0.81)^1 + (0.19)(0.81)^2 \approx 0.469$

(c) $P(x > 3) = 1 - P(x \le 3) = 1 - 0.469 = 0.531$

19. $p = 0.002$

(a) $P(10) = (0.002)(0.998)^9 \approx 0.002$

(b) $P(\text{1st, 2nd, or 3rd part is defective}) = P(1) + P(2) + P(3)$

$$= (0.002)(0.998)^0 + (0.002)(0.998)^1 + (0.002)(0.998)^2 \approx 0.006$$

(c) $P(x > 10) = 1 - P(x \leq 10) = 1 - [P(1) + P(2) + \cdots + P(10)] = 1 - [0.020] \approx 0.980$

21. $\mu = 3$

(a) $P(5) = \dfrac{3^5 e^{-3}}{5!} \approx 0.101$

(b) $P(x \geq 5) = 1 - (P(0) + P(1) + P(2) + P(3) + P(4))$

$$\approx 1 - (0.050 + 0.149 + 0.224 + 0.224 + 0.168)$$

$$= 0.185$$

(c) $P(x > 5) = 1 - (P(0) + P(1) + P(2) + P(3) + P(4) + P(5))$

$$\approx 1 - (0.050 + 0.149 + 0.224 + 0.224 + 0.168 + 0.101)$$

$$= 0.084$$

23. $\mu = 0.6$

(a) $P(1) = 0.329$

(b) $P(x \leq 1) = P(0) + P(1) = 0.549 + 0.329 = 0.878$

(c) $P(x > 1) = 1 - P(x \leq 1) = 1 - 0.878 = 0.122$

25. (a) $n = 6000, \ p = 0.0004$

$$P(4) = \frac{6000!}{5996!4!}(0.0004)^4(0.9996)^{5996} \approx 0.1254235482$$

(b) $\mu = \dfrac{6000}{2500} = 2.4$ warped glass items per 2500.

$P(4) = 0.1254084986$

The results are approximately the same.

27. $p = 0.001$

(a) $\mu = \dfrac{1}{p} = \dfrac{1}{0.001} = 1000$

$\sigma^2 = \dfrac{q}{p^2} = \dfrac{0.999}{(0.001)^2} = 999{,}000$

$\sigma = \sqrt{\sigma^2} \approx 999.5$

On average you would have to play 1000 times until you won the lottery. The standard deviation is 999.5 times.

(b) 1000 times

Lose money. On average you would win $500 every 1000 times you play the lottery. So, the net gain would be $-\$500$.

29. $\mu = 3.8$

(a) $\sigma^2 = 3.8$

$\sigma = \sqrt{\sigma^2} \approx 1.9$

The standard deviation is 1.9 strokes.

(b) $P(x > 72) = 1 - P(x \le 72) = 1 - 0.695 = 0.305$

CHAPTER 4 REVIEW EXERCISE SOLUTIONS

1. Discrete

2. Continuous

3. Continuous

4. Discrete

5. No, $\Sigma P(x) \ne 1$.

6. Yes

7. Yes

8. No, $P(5) > 1$ and $\Sigma P(x) \ne 1$.

9. Yes

10. No, $\Sigma P(x) \ne 1$.

11. (a)

x	Frequency	$P(x)$	$xP(x)$	$x - \mu$	$(x - \mu)^2$	$(x - \mu)^2P(x)$
2	3	0.005	0.009	−4.371	19.104	0.088
3	12	0.018	0.055	−3.371	11.362	0.210
4	72	0.111	0.443	−2.371	5.621	0.623
5	115	0.177	0.885	−1.371	1.879	0.332
6	169	0.260	1.560	−0.371	0.137	0.036
7	120	0.185	1.292	0.629	0.396	0.073
8	83	0.128	1.022	1.629	2.654	0.339
9	48	0.074	0.665	2.629	6.913	0.510
10	22	0.034	0.338	3.629	13.171	0.446
11	6	0.009	0.102	4.629	21.430	0.198
	$n = 650$	$\sum P(x) = 1$	$\sum xP(x) = 6.371$			$\sum (x - \mu)^2P(x) = 2.855$

(b) **Pages per Section**

(c) $\mu = \Sigma xP(x) \approx 6.4$

$\sigma^2 = \Sigma(x - \mu)^2P(x) \approx 2.9$

$\sigma = \sqrt{\sigma^2} \approx 1.7$

13. (a)

x	Frequency	$P(x)$	$xP(x)$	$x - \mu$	$(x - \mu)^2$	$(x - \mu)^2P(x)$
0	3	0.015	0.000	−2.315	5.359	0.080
1	38	0.190	0.190	−1.315	1.729	0.329
2	83	0.415	0.830	−0.315	0.099	0.041
3	52	0.260	0.780	0.685	0.469	0.122
4	18	0.090	0.360	1.685	2.839	0.256
5	5	0.025	0.125	2.685	7.209	0.180
6	1	0.005	0.030	3.685	13.579	0.068
	$n = 200$	$\sum P(x) = 1$	$\sum xP(x) = 2.315$			$\sum (x - \mu)^2P(x) = 1.076$

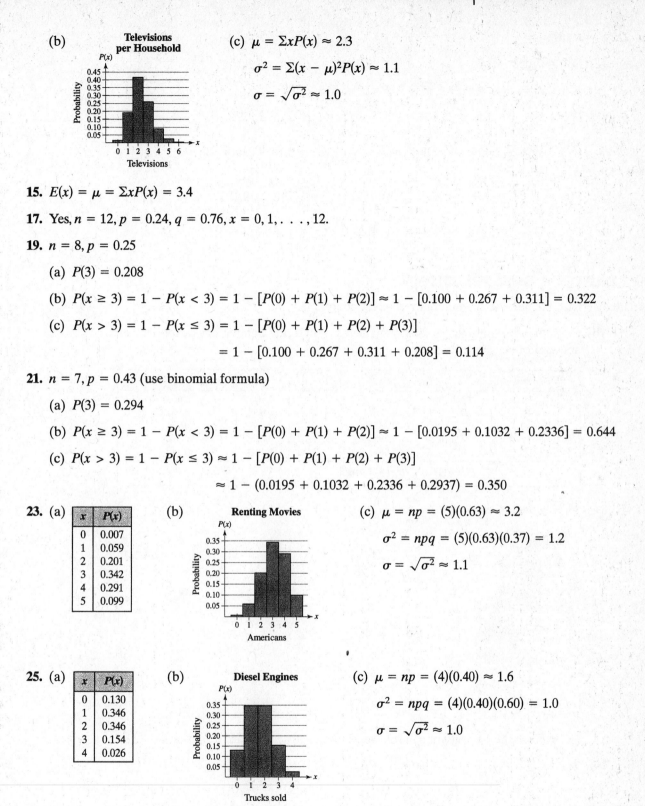

(b)

Televisions per Household

(c) $\mu = \Sigma x P(x) \approx 2.3$

$\sigma^2 = \Sigma(x - \mu)^2 P(x) \approx 1.1$

$\sigma = \sqrt{\sigma^2} \approx 1.0$

15. $E(x) = \mu = \Sigma x P(x) = 3.4$

17. Yes, $n = 12, p = 0.24, q = 0.76, x = 0, 1, \ldots, 12$.

19. $n = 8, p = 0.25$

(a) $P(3) = 0.208$

(b) $P(x \geq 3) = 1 - P(x < 3) = 1 - [P(0) + P(1) + P(2)] \approx 1 - [0.100 + 0.267 + 0.311] = 0.322$

(c) $P(x > 3) = 1 - P(x \leq 3) = 1 - [P(0) + P(1) + P(2) + P(3)]$

$= 1 - [0.100 + 0.267 + 0.311 + 0.208] = 0.114$

21. $n = 7, p = 0.43$ (use binomial formula)

(a) $P(3) = 0.294$

(b) $P(x \geq 3) = 1 - P(x < 3) = 1 - [P(0) + P(1) + P(2)] \approx 1 - [0.0195 + 0.1032 + 0.2336] = 0.644$

(c) $P(x > 3) = 1 - P(x \leq 3) \approx 1 - [P(0) + P(1) + P(2) + P(3)]$

$\approx 1 - (0.0195 + 0.1032 + 0.2336 + 0.2937) = 0.350$

23. (a)

x	$P(x)$
0	0.007
1	0.059
2	0.201
3	0.342
4	0.291
5	0.099

(b) **Renting Movies**

(c) $\mu = np = (5)(0.63) \approx 3.2$

$\sigma^2 = npq = (5)(0.63)(0.37) = 1.2$

$\sigma = \sqrt{\sigma^2} \approx 1.1$

25. (a)

x	$P(x)$
0	0.130
1	0.346
2	0.346
3	0.154
4	0.026

(b) **Diesel Engines**

(c) $\mu = np = (4)(0.40) \approx 1.6$

$\sigma^2 = npq = (4)(0.40)(0.60) = 1.0$

$\sigma = \sqrt{\sigma^2} \approx 1.0$

27. $p = 0.167$

    (a) $P(4) \approx 0.096$

    (b) $P(x \le 4) = P(1) + P(2) + P(3) + P(4) \approx 0.518$

    (c) $P(x > 3) = 1 - P(x \le 3) = 1 - [P(1) + P(2) + P(3)] \approx 0.579$

29. $\mu = \dfrac{2457}{36} \approx 68.25$ lightning deaths/year $\rightarrow \mu = \dfrac{68.25}{365} \approx 0.1869$ deaths/day

    (a) $P(0) = \dfrac{0.1869^0 e^{0.1869}}{0!} \approx 0.830$ (b) $P(1) = \dfrac{0.1869^1 e^{1869}}{1!} \approx 0.155$

    (c) $P(x > 1) = 1 - [P(0) + P(1)] = 1 - [0.830 + 0.155] = 0.015$

CHAPTER 4 QUIZ SOLUTIONS

1. (a) Discrete because the random variable is countable.

    (b) Continuous because the random variable has an infinite number of possible outcomes and cannot be counted.

2. (a)

x	Frequency	$P(x)$	$xP(x)$	$x - \mu$	$(x - \mu)^2$	$(x - \mu)^2 P(x)$
1	70	0.398	0.398	-1.08	1.166	0.464
2	41	0.233	0.466	-0.08	0.006	0.001
3	49	0.278	0.835	0.92	0.846	0.236
4	13	0.074	0.295	1.92	3.686	0.272
5	3	0.017	0.085	2.92	8.526	0.145
	$n = 176$	$\sum P(x) = 1$	$\sum xP(x) = 2.080$			$\sum (x - \mu)^2 P(x) = 1.119$

    (b)

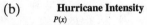
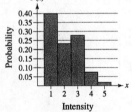

Hurricane Intensity

    (c) $\mu = \Sigma xP(x) \approx 2.1$

        $\sigma^2 = \Sigma (x - \mu)^2 P(x) \approx 1.1$

        $\sigma = \sqrt{\sigma^2} \approx 1.1$

    On average the intensity of a hurricane will be 2.1. The standard deviation is 1.1.

    (d) $P(x \ge 4) = P(4) + P(5) = 0.074 + 0.017 = 0.091$

3. $n = 8, p = 0.80$

(a)

x	$P(x)$
0	0.000003
1	0.000082
2	0.001147
3	0.009175
4	0.045875
5	0.146801
6	0.293601
7	0.335544
8	0.167772

(b)

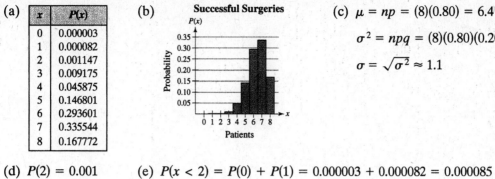

(c) $\mu = np = (8)(0.80) = 6.4$

$\sigma^2 = npq = (8)(0.80)(0.20) = 1.3$

$\sigma = \sqrt{\sigma^2} \approx 1.1$

(d) $P(2) = 0.001$

(e) $P(x < 2) = P(0) + P(1) = 0.000003 + 0.000082 = 0.000085$

4. $\mu = 5$

(a) $P(5) = 0.176$

(b) $P(x < 5) = P(0) + P(1) + P(2) + P(3) + P(4)$

$= 0.00674 + 0.03369 + 0.08422 + 0.14037 + 0.17547 \approx 0.440$

(c) $P(0) \approx 0.007$

Normal Probability Distributions

5.1 INTRODUCTION TO NORMAL DISTRIBUTIONS AND THE STANDARD NORMAL DISTRIBUTION

5.1 Try It Yourself Solutions

1a. A: 45, B: 60, C: 45 (B has the greatest mean.)

b. Curve C is more spread out, so curve C has the greatest standard deviation.

2a. Mean = 3.5 feet

b. Inflection points: 3.3 and 3.7
Standard deviation = 0.2 foot

3a. (1) 0.0143   (2) 0.9850

4a.

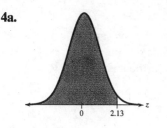

b. 0.9834

5a.

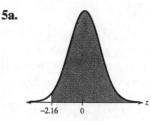

b. 0.0154

c. Area = 1 − 0.0154 = 0.9846

6a. 0.0885   **b.** 0.0154

c. Area = 0.0885 − 0.0154 = 0.0731

5.1 EXERCISE SOLUTIONS

1. Answers will vary.

3. Answers will vary.
Similarities: Both curves will have the same line of symmetry.
Differences: One curve will be more spread out than the other.

5. $\mu = 0$, $\sigma = 1$

7. "The" standard normal distribution is used to describe one specific normal distribution ($\mu = 0$, $\sigma = 1$). "A" normal distribution is used to describe a normal distribution with any mean and standard deviation.

9. No, the graph crosses the x-axis.

11. Yes, the graph fulfills the properties of the normal distribution.

13. No, the graph is skewed to the right.

15. The histogram represents data from a normal distribution because it's bell-shaped.

17. (Area left of $z = 1.2$) − (Area left of $z = 0$) = 0.8849 − 0.5 = 0.3849

19. (Area left of $z = 1.5$) − (Area left of $z = -0.5$) = 0.9332 − 0.3085 = 0.6247

21. 0.9131   **23.** 0.975

25. 1 − 0.2578 = 0.7422   **27.** 1 − 0.8997 = 0.1003

29. 0.005   **31.** 1 − 0.9469 = 0.0531

33. 0.9382 − 0.5 = 0.4382   **35.** 0.5 − 0.0630 = 0.437

37. 0.9750 − 0.0250 = 0.95   **39.** 0.1003 + 0.1003 = 0.2006

41. (a)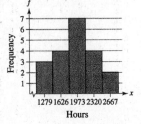

It is reasonable to assume that the life span is normally distributed because the histogram is nearly symmetric and bell-shaped.

(b) $\bar{x} = 1941.35$ $s \approx 432.385$

(c) The sample mean of 1941.35 hours is less than the claimed mean, so on the average the bulbs in the sample lasted for a shorter time. The sample standard deviation of 432 hours is greater than the claimed standard deviation, so the bulbs in the sample had a greater variation in life span than the manufacturer's claim.

43. (a) A = 92.994   B = 93.004   C = 93.014   D = 93.018

(b) $x = 93.014 \Longrightarrow z = \dfrac{x - \mu}{\sigma} = \dfrac{93.014 - 93.01}{0.005} = 0.8$

$x = 93.018 \Longrightarrow z = \dfrac{x - \mu}{\sigma} = \dfrac{93.018 - 93.01}{0.005} = 1.6$

$x = 93.004 \Longrightarrow z = \dfrac{x - \mu}{\sigma} = \dfrac{93.004 - 93.01}{0.005} = -1.2$

$x = 92.994 \Longrightarrow z = \dfrac{x - \mu}{\sigma} = \dfrac{92.994 - 93.01}{0.005} = -3.2$

(c) $x = 92.994$ is unusual due to a relatively small z-score (-3.2).

45. (a) A = 1186   B = 1406   C = 1848   D = 2177

(b) $x = 1406 \Longrightarrow z = \dfrac{x - \mu}{\sigma} = \dfrac{1406 - 1518}{308} = -0.36$

$x = 1848 \Longrightarrow z = \dfrac{x - \mu}{\sigma} = \dfrac{1848 - 1518}{308} = 1.07$

$x = 2177 \Longrightarrow z = \dfrac{x - \mu}{\sigma} = \dfrac{2177 - 1518}{308} = 2.14$

$x = 1186 \Longrightarrow z = \dfrac{x - \mu}{\sigma} = \dfrac{1186 - 1518}{308} = -1.08$

(c) $x = 2177$ is unusual due to a relatively large z-score (2.14).

47. 0.6915

49. $1 - 0.95 = 0.05$

51. $0.8413 - 0.3085 = 0.5328$

53. $P(z < 1.45) = 0.9265$

55. $P(z > -0.95) = 1 - P(z < -0.95) = 1 - 0.1711 = 0.8289$

57. $P(-0.89 < z < 0) = 0.5 - 0.1867 = 0.3133$

59. $P(-1.65 < z < 1.65) = 0.9505 - 0.0495 = 0.901$

61. $P(z < -2.58 \text{ or } z > 2.58) = 2(0.0049) = 0.0098$

63.

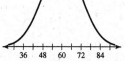

The normal distribution curve is centered at its mean (60) and has 2 points of inflection (48 and 72) representing $\mu \pm \sigma$.

65. (a) Area under curve $=$ area of rectangle $=$ (base)(height) $= (1)(1) = 1$

    (b) $P(0.25 < x < 0.5) =$ (base)(height) $= (0.25)(1) = 0.25$

    (c) $P(0.3 < x < 0.7) =$ (base)(height) $= (0.4)(1) = 0.4$

5.2 NORMAL DISTRIBUTIONS: FINDIING PROBABILITIES

5.2 Try It Yourself Solutions

1a.

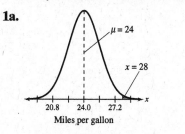

Miles per gallon

b. $z = \dfrac{x - \mu}{\sigma} = \dfrac{28 - 24}{1.6} = 2.50$

c. $P(z < 2.50) = 0.9938$

    $P(z > 2.50) = 1 - 0.9938 = 0.0062$

d. The probability that a randomly selected manual transmission Focus will get more than 28 mpg in city driving is 0.0062.

2a.

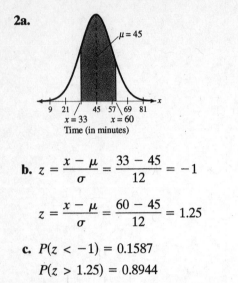

b. $z = \dfrac{x - \mu}{\sigma} = \dfrac{33 - 45}{12} = -1$

$z = \dfrac{x - \mu}{\sigma} = \dfrac{60 - 45}{12} = 1.25$

c. $P(z < -1) = 0.1587$

$P(z > 1.25) = 0.8944$

$0.8944 - 0.1587 = 0.7357$

d. If 150 shoppers enter the store, then you would expect $150(0.7357) \approx 110$ shoppers to be in the store between 33 and 60 minutes.

3a. Read user's guide for the technology tool.

b. Enter the data.

c. $P(190 < x < 225) = P(-1 < z < 0.4) = 0.4968$

The probability that a randomly selected U.S. man's cholesterol is between 190 and 225 is about 0.4968.

5.2 EXERCISE SOLUTIONS

1. $P(x < 80) = P(z < -1.2) = 0.1151$

3. $P(x > 92) = P(z > 1.2) = 1 - 0.8849 = 0.1151$

5. $P(70 < x < 80) = P(-3.2 < z < -1.2) = 0.1151 - 0.0007 = 0.1144$

7. $P(200 < x < 450) = P(-2.68 < z < -0.47) = 0.3192 - 0.0037 = 0.3155$

9. $P(200 < x < 239) = P(0.39 < z < 1.48) = 0.9306 - 0.6517 = 0.2789$

11. $P(141 < x < 151) = P(0.84 < z < 2.94) = 0.9984 - 0.7995 = 0.1989$

13. (a) $P(x < 66) = P(z < -1.2) = 0.1151$

(b) $P(66 < x < 72) = P(-1.2 < z < 0.8) = 0.7881 - 0.1151 = 0.6730$

(c) $P(x > 72) = P(z > 0.8) = 1 - P(z < 0.8) = 1 - 0.7881 = 0.2119$

15. (a) $P(x < 17) = P(z < -1.67) = 0.0475$

(b) $P(20 < x < 29) = P(-0.98 < z < 1.12) = 0.8686 - 0.1635 = 0.7051$

(c) $P(x > 32) = P(z > 1.81) = 1 - P(z < 1.81) = 1 - 0.9649 = 0.0351$

17. (a) $P(x < 5) = P(z < -2) = 0.0228$

(b) $P(5.5 < x < 9.5) = P(-1.5 < z < 2.5) = 0.9938 - 0.0668 = 0.927$

(c) $P(x > 10) = P(z > 3) = 1 - P(z < 3) = 1 - 0.9987 = 0.0013$

19. (a) $P(x < 4) = (z < -2.44) = 0.0073$

(b) $P(5 < x < 7) = P(-1.33 < z < 0.89) = 0.8133 - 0.0918 = 0.7215$

(c) $P(x > 8) = P(z > 2) = 1 - 0.9772 = 0.0228$

21. (a) $P(x < 600) = P(z < 0.86) = 0.8051 \Rightarrow 80.51\%$

(b) $P(x > 550) = P(z > 0.42) = 1 - P(z < 0.42) = 1 - 0.6628 = 0.3372$
$(1000)(0.3372) = 337.2 \Rightarrow 337$ scores

23. (a) $P(x < 200) = P(z < 0.39) = 0.6517 \Rightarrow 65.17\%$

(b) $P(x > 240) = P(z > 1.51) = 1 - P(z < 1.51) = 1 - 0.9345 = 0.0655$
$(250)(0.0655) = 16.375 \Rightarrow 16$ women

25. (a) $P(x > 11) = P(z > 0.5) = 1 - P(z < 0.5) = 1 - 0.6915 = 0.3085 \Rightarrow 30.85\%$

(b) $P(x < 8) = P(z < -1) = 0.1587$
$(200)(0.1587) = 31.74 \Rightarrow 32$ fish

27. (a) $P(x > 4) = P(z > -3) = 1 - P(z < -3) = 1 - 0.0013 = 0.9987 \Rightarrow 99.87\%$

(b) $P(x < 5) = P(z < -2) = 0.0228$
$(35)(0.0228) = 0.798 \Rightarrow 1$ adult

29. $P(x > 2065) = P(z > 2.17) = 1 - P(z < 2.17) = 1 - 0.9850 = 0.0150 \Rightarrow 1.5\%$

It is unusual for a battery to have a life span that is more than 2065 hours because of the relatively large z-score (2.17).

31. Out of control, because the 10th observation plotted beyond 3 standard deviations.

33. Out of control, because the first nine observations lie below the mean and since two out of three consecutive points lie more than 2 standard deviations from the mean.

5.3 NORMAL DISTRIBUTIONS: FINDING VALUES

5.3 Try It Yourself Solutions

1ab. (1)   (2)

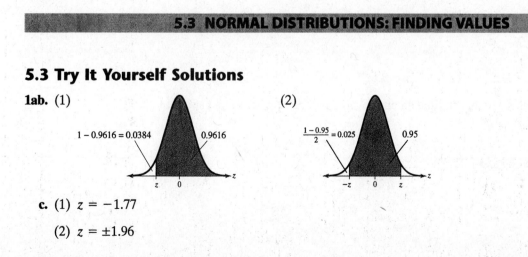

c. (1) $z = -1.77$

(2) $z = \pm 1.96$

2a. (1)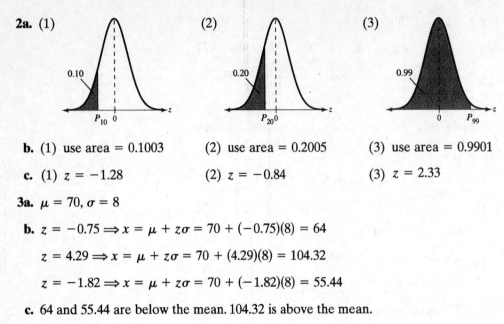

(2)

(3)

b. (1) use area = 0.1003   (2) use area = 0.2005   (3) use area = 0.9901

c. (1) $z = -1.28$ (2) $z = -0.84$ (3) $z = 2.33$

3a. $\mu = 70$, $\sigma = 8$

b. $z = -0.75 \Rightarrow x = \mu + z\sigma = 70 + (-0.75)(8) = 64$

     $z = 4.29 \Rightarrow x = \mu + z\sigma = 70 + (4.29)(8) = 104.32$

     $z = -1.82 \Rightarrow x = \mu + z\sigma = 70 + (-1.82)(8) = 55.44$

c. 64 and 55.44 are below the mean. 104.32 is above the mean.

4a.

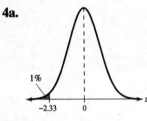

b. $z = -2.33$

c. $x = \mu + z\sigma = 142 + (-2.33)(6.51) \approx 126.83$

d. So, the longest braking distance a Honda Accord could have and still be in the top 1% is 127 feet.

5a.

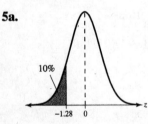

b. $z = -1.28$

c. $x = \mu + z\sigma = 11.2 + (-1.28)(2.1) = 8.512$

d. So, the maximum length of time an employee could have worked and still be laid off is 8 years.

5.3 EXERCISE SOLUTIONS

1. $z = -0.70$ **3.** $z = 0.34$

5. $z = -0.16$ **7.** $z = 2.39$

9. $z = -1.645$ **11.** $z = 1.555$

13. $z = -2.33$ **15.** $z = -0.84$

17. $z = 1.175$ **19.** $z = -0.67$

21. $z = 0.67$ **23.** $z = -0.39$

25. $z = -0.38$ **27.** $z = -0.58$

29. $z = \pm 1.645$

31. 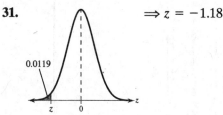 $\Rightarrow z = -1.18$ **33.** 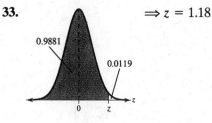 $\Rightarrow z = 1.18$

35. 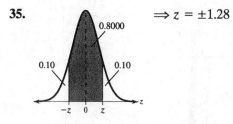 $\Rightarrow z = \pm 1.28$ **37.** 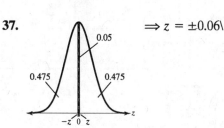 $\Rightarrow z = \pm 0.06\backslash$

39. (a) 95th percentile $\Rightarrow$ Area $= 0.95 \Rightarrow z = 1.645$

$x = \mu + z\sigma = 64.1 + (1.645)(2.71) = 68.56$ inches

(b) 1st quartile $\Rightarrow$ Area $= 0.25 \Rightarrow z = -0.67$

$x = \mu + z\sigma = 64.1 + (-0.67)(2.71) = 62.28$ inches

41. (a) 10th percentile $\Rightarrow$ Area $= 0.10 \Rightarrow z = -1.28$

$x = \mu + z\sigma = 17.1 + (-1.28)(4) = 11.98$ pounds

(b) 3rd quartile $\Rightarrow$ Area $= 0.75 \Rightarrow z = 0.67$

$x = \mu + z\sigma = 17.1 + (0.67)(4) = 19.78$ pounds

43. (a) Top 30% $\Rightarrow$ Area $= 0.70 \Rightarrow z = 0.52$

$x = \mu + z\sigma = 127 + (0.52)(23.5) = 139.22$ days

(b) Bottom 10% $\Rightarrow$ Area $= 0.10 \Rightarrow z = -1.28$

$x = \mu + z\sigma = 127 + (-1.28)(23.5) = 96.92$ days

45. Lower 5% $\Rightarrow$ Area $= 0.05 \Rightarrow z = -1.645$

$x = \mu + z\sigma = 20 + (-1.645)(0.07) = 19.88$

47. Bottom 10% $\Rightarrow$ Area $= 0.10 \Rightarrow z = -1.28$

$x = \mu + z\sigma = 30,000 + (-1.28)(2500) = 26,800$

Tires which wear out by 26,800 miles will be replaced free of charge.

49. Top 1% $\Rightarrow$ Area $= 0.99 \Rightarrow z = 2.33$

$x = \mu + z\sigma \Rightarrow 8 = \mu + (2.33)(0.03) \Rightarrow \mu = 7.930$ ounces

5.4 SAMPLING DISTRIBUTIONS AND THE CENTRAL LIMIT THEOREM

5.4 Try It Yourself Solutions

1a.

Sample	Mean	Sample	Mean	Sample	Mean	Sample	Mean
1, 1, 1	1	3, 1, 1	1.67	5, 1, 1	2.33	7, 1, 1	3
1, 1, 3	1.67	3, 1, 3	2.33	5, 1, 3	3	7, 1, 3	3.67
1, 1, 5	2.33	3, 1, 5	3	5, 1, 5	3.67	7, 1, 5	4.33
1, 1, 7	3	3, 1, 7	3.67	5, 1, 7	4.33	7, 1, 7	5
1, 3, 1	1.67	3, 3, 1	2.33	5, 3, 1	3	7, 3, 1	3.67
1, 3, 3	2.33	3, 3, 3	3	5, 3, 3	3.67	7, 3, 3	4.33
1, 3, 5	3	3, 3, 5	3.67	5, 3, 5	4.33	7, 3, 5	5
1, 3, 7	3.67	3, 3, 7	4.33	5, 3, 7	5	7, 3, 7	5.67
1, 5, 1	2.33	3, 5, 1	3	5, 5, 1	3.67	7, 5, 1	4.33
1, 5, 3	3	3, 5, 3	3.67	5, 5, 3	4.33	7, 5, 3	5
1, 5, 5	3.67	3, 5, 5	4.33	5, 5, 5	5	7, 5, 5	5.67
1, 5, 7	4.33	3, 5, 7	5	5, 5, 7	5.67	7, 5, 7	6.33
1, 7, 1	3	3, 7, 1	3.67	5, 7, 1	4.33	7, 7, 1	5
1, 7, 3	3.67	3, 7, 3	4.33	5, 7, 3	5	7, 7, 3	5.67
1, 7, 5	4.33	3, 7, 5	5	5, 7, 5	5.67	7, 7, 5	6.33
1, 7, 7	5	3, 7, 7	5.67	5, 7, 7	6.33	7, 7, 7	7

b.

$\bar{x}$	f	Probability
1	1	0.0156
1.67	3	0.0469
2.33	6	0.0938
3	10	0.1563
3.67	12	0.1875
4.33	12	0.1875
5	10	0.1563
5.67	6	0.0938
6.33	3	0.0469
7	1	0.0156

c. $\mu_{\bar{x}} = \mu = 4,$

$$\sigma_{\bar{x}}^2 = \frac{\sigma^2}{n} = \frac{5}{3} = 1.667,$$

$$\sigma_{\bar{x}} = \frac{\sigma}{\sqrt{n}} = \frac{\sqrt{5}}{\sqrt{3}} = 1.291$$

$\mu_{\bar{x}} = 4, \sigma_{\bar{x}}^2 \approx 1.667, \sigma_{\bar{x}} \approx 1.291$

2a. $\mu_{\bar{x}} = \mu = 64, \sigma_{\bar{x}} = \frac{\sigma}{\sqrt{n}} = \frac{9}{\sqrt{100}} = 0.9$

b. $n = 100$

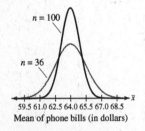

59.5 61.0 62.5 64.0 65.5 67.0 68.5
Mean of phone bills (in dollars)

c. With a larger sample size, the mean stays the same but the standard deviation decreases.

3a. $\mu_{\bar{x}} = \mu = 3.5, \sigma_{\bar{x}} = \frac{\sigma}{\sqrt{n}} = \frac{0.2}{\sqrt{16}} = 0.05$

b.

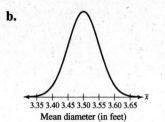

3.35 3.40 3.45 3.50 3.55 3.60 3.65
Mean diameter (in feet)

4a. $\mu_{\bar{x}} = \mu = 25, \sigma_{\bar{x}} = \dfrac{\sigma}{\sqrt{n}} = \dfrac{1.5}{\sqrt{100}} = 0.15$

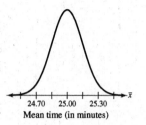

24.70   25.00   25.30
Mean time (in minutes)

b. $\bar{x} = 24.7$: $z = \dfrac{\bar{x} - \mu}{\dfrac{\sigma}{\sqrt{n}}} = \dfrac{24.7 - 25}{\dfrac{1.5}{\sqrt{100}}} = -\dfrac{0.3}{0.15} = -2$

$\bar{x} = 25.5$: $z = \dfrac{\bar{x} - \mu}{\dfrac{\sigma}{\sqrt{n}}} = \dfrac{25.5 - 25}{\dfrac{1.5}{\sqrt{100}}} = \dfrac{0.5}{0.15} = 3.33$

c. $P(z < -2) \approx 0.0228$

$P(z < 3.33) = 0.9996$

$P(8.7 < \bar{x} < 9.5) = P(-2 < z < 3.33) = 0.9996 - 0.0228 = 0.9768$

5a. $\mu_{\bar{x}} = \mu = 306{,}258, \sigma_{\bar{x}} = \dfrac{\sigma}{\sqrt{n}} = \dfrac{44{,}000}{\sqrt{12}} \approx 12{,}701.7$

280,855   306,258   331,661
Mean sales price (in dollars)

b. $\bar{x} = 280{,}000$: $z = \dfrac{\bar{x} - \mu}{\dfrac{\sigma}{\sqrt{n}}} = \dfrac{280{,}000 - 306{,}258}{\dfrac{44{,}000}{\sqrt{12}}} = \dfrac{-26{,}258}{12{,}701.7} = -2.07$

c. $P(\bar{x} > 200{,}000) = P(z > -2.07) = 1 - P(z < -2.07) = 1 - 0.0192 = 0.9808$

6a. $x = 700$: $z = \dfrac{x - \mu}{\sigma} = \dfrac{700 - 625}{150} = 0.5$

$\bar{x} = 700$: $z = \dfrac{\bar{x} - \mu}{\dfrac{\sigma}{\sqrt{n}}} = \dfrac{700 - 625}{\dfrac{150}{\sqrt{10}}} = \dfrac{75}{47.43} = 1.58$

b. $P(z < 0.5) = 0.6915$

$P(z < 1.58) = 0.9429$

c. There is a 69% chance an individual receiver will cost less than \$700. There is a 94% chance that the mean of a sample of 10 receivers is less than \$700.

5.4 EXERCISE SOLUTIONS

1. $\mu_{\bar{x}} = \mu = 100$

$\sigma_{\bar{x}} = \dfrac{\sigma}{\sqrt{n}} = \dfrac{15}{\sqrt{50}} = 2.121$

3. $\mu_{\bar{x}} = \mu = 100$

$\sigma_{\bar{x}} = \dfrac{\sigma}{\sqrt{n}} = \dfrac{15}{\sqrt{250}} = 0.949$

5. False. As the size of the sample increases, the mean of the distribution of the sample mean does not change.

7. False. The shape of the sampling distribution of sample means is normal for large sample sizes even if the shape of the population is non-normal.

9. $\mu_{\bar{x}} = 3.5$, $\sigma_{\bar{x}} = 1.708$

$\mu = 3.5$, $\sigma = 2.958$

The means are equal but the standard deviation of the sampling distribution is smaller.

Sample	Mean	Sample	Mean	Sample	Mean	Sample	Mean
0, 0, 0	0	2, 0, 0	0.67	4, 0, 0	1.33	8, 0, 0	2.67
0, 0, 2	0.67	2, 0, 2	1.33	4, 0, 2	2	8, 0, 2	3.33
0, 0, 4	1.33	2, 0, 4	2	4, 0, 4	2.67	8, 0, 4	4
0, 0, 8	2.67	2, 0, 8	3.33	4, 0, 8	4	8, 0, 8	5.33
0, 2, 0	0.67	2, 2, 0	1.33	4, 2, 0	2	8, 2, 0	3.33
0, 2, 2	1.33	2, 2, 2	2	4, 2, 2	2.67	8, 2, 2	4
0, 2, 4	2	2, 2, 4	2.67	4, 2, 4	3.33	8, 2, 4	4.67
0, 2, 8	3.33	2, 2, 8	4	4, 2, 8	4.67	8, 2, 8	6
0, 4, 0	1.33	2, 4, 0	2	4, 4, 0	2.67	8, 4, 0	4
0, 4, 2	2	2, 4, 2	2.67	4, 4, 2	3.33	8, 4, 2	4.67
0, 4, 4	2.67	2, 4, 4	3.33	4, 4, 4	4	8, 4, 4	5.33
0, 4, 8	4	2, 4, 8	4.67	4, 4, 8	5.33	8, 4, 8	6.67
0, 8, 0	2.67	2, 8, 0	3.33	4, 8, 0	4	8, 8, 0	5.33
0, 8, 2	3.33	2, 8, 2	4	4, 8, 2	4.67	8, 8, 2	6
0, 8, 4	4	2, 8, 4	4.67	4, 8, 4	5.33	8, 8, 4	6.67
0, 8, 8	5.33	2, 8, 8	6	4, 8, 8	6.67	8, 8, 8	8

11. (c) Because $\mu_{\bar{x}} = 16.5$, $\sigma_{\bar{x}} = \dfrac{\sigma}{\sqrt{n}} = \dfrac{11.9}{\sqrt{100}} = 1.19$ and the graph approximates a normal curve.

13. $z = \dfrac{\bar{x} - \mu}{\dfrac{\sigma}{\sqrt{n}}} = \dfrac{12.2 - 12}{\dfrac{0.95}{\sqrt{36}}} = \dfrac{0.2}{0.158} = 1.26$

$P(\bar{x} < 12.2) = P(z < 1.26) = 0.8962$

15. $z = \dfrac{\bar{x} - \mu}{\dfrac{\sigma}{\sqrt{n}}} = \dfrac{221 - 220}{\dfrac{39}{\sqrt{75}}} = \dfrac{1}{0.450} = 2.22$

$P(\bar{x} > 221) = P(z > 2.22) = 1 - P(z < 2.22) = 1 - 0.9868 = 0.0132$

The probability is unusual because it is less than 0.05.

17. $\mu_{\bar{x}} = 87.5$

$\sigma_{\bar{x}} = \dfrac{\sigma}{\sqrt{n}} = \dfrac{6.25}{\sqrt{12}} \approx 1.804$

19. $\mu_{\bar{x}} = 224$

$\sigma_{\bar{x}} = \dfrac{\sigma}{\sqrt{n}} = \dfrac{8}{\sqrt{40}} = 1.265$

82.1 83.9 85.7 87.5 89.3 91.1 92.9
Mean height (in feet)

221.5 224 226.5
Mean price (in dollars)

21. $\mu_{\bar{x}} = 110$

$\sigma_{\bar{x}} = \dfrac{\sigma}{\sqrt{n}} = \dfrac{38.5}{\sqrt{20}} \approx 8.609$

92.8 110 127.2
Mean consumption of
red meat (in pounds)

23. $\mu_{\bar{x}} = 87.5, \sigma_{\bar{x}} = \dfrac{\sigma}{\sqrt{n}} = \dfrac{6.25}{\sqrt{24}} \approx 1.276$

$\mu_{\bar{x}} = 87.5, \sigma_{\bar{x}} = \dfrac{\sigma}{\sqrt{n}} = \dfrac{6.25}{\sqrt{36}} \approx 1.042$

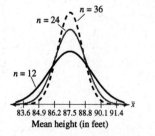

$n = 36$
$n = 24$
$n = 12$

83.6 84.9 86.2 87.5 88.8 90.1 91.4
Mean height (in feet)

As the sample size increases, the standard error decreases, while the mean of the sample means remains constant.

25. $z = \dfrac{\bar{x} - \mu}{\dfrac{\sigma}{\sqrt{n}}} = \dfrac{44{,}000 - 46{,}700}{\dfrac{5600}{\sqrt{42}}} = \dfrac{-2700}{864.10} \approx -3.12$

$P(\bar{x} < 44{,}000) = P(z < -3.12) = 0.0009$

27. $z = \dfrac{\bar{x} - \mu}{\dfrac{\sigma}{\sqrt{n}}} = \dfrac{2.768 - 2.818}{\dfrac{0.045}{\sqrt{32}}} = \dfrac{-0.05}{0.00795} \approx -6.29$

$z = \dfrac{\bar{x} - \mu}{\dfrac{\sigma}{\sqrt{n}}} = \dfrac{2.918 - 2.818}{\dfrac{0.045}{\sqrt{32}}} = \dfrac{0.1}{0.00795} \approx 12.57$

$P(2.768 < \bar{x} < 2.918) = P(-6.29 < z < 12.57) \approx 1 - 0 = 1$

29. $z = \dfrac{\bar{x} - \mu}{\dfrac{\sigma}{\sqrt{n}}} = \dfrac{66 - 64.1}{\dfrac{2.71}{\sqrt{60}}} = \dfrac{1.9}{0.350} \approx 5.43$

$P(\bar{x} > 66) = P(z > 5.43) \approx 0$

31. $z = \dfrac{\bar{x} - \mu}{\sigma} = \dfrac{70 - 64.1}{2.71} \approx 2.18$

$P(x < 70) = P(z < 2.18) = 0.9854$

$z = \dfrac{\bar{x} - \mu}{\dfrac{\sigma}{\sqrt{n}}} = \dfrac{70 - 64.1}{\dfrac{2.71}{\sqrt{20}}} = \dfrac{5.9}{0.606} \approx 9.73$

$P(\bar{x} < 70) = P(z < 9.73) \approx 1$

It is more likely to select a sample of 20 women with a mean height less than 70 inches, because the sample of 20 has a higher probability.

33. $z = \dfrac{\bar{x} - \mu}{\dfrac{\sigma}{\sqrt{n}}} = \dfrac{127.9 - 128}{\dfrac{0.20}{\sqrt{40}}} = \dfrac{-0.1}{0.032} \approx 3.16$

$P(\bar{x} < 127.9) = P(z < -3.16) = 0.0008$

Yes, it is very unlikely that we would have randomly sampled 40 cans with a mean equal to 127.9 ounces, because it is more than 2 standard deviations from the mean of the sample means.

35. (a) $\mu = 96$

$\sigma = 0.5$

$z = \dfrac{\bar{x} - \mu}{\dfrac{\sigma}{\sqrt{\mu}}} = \dfrac{96.25 - 96}{\dfrac{0.5}{\sqrt{90}}} = \dfrac{0.25}{0.079} \approx 3.16$

$P(\bar{x} \geq 96.25) = P(z > 3.16) = 1 - P(z < 3.16) = 1 - 0.9992 = 0.0008$

(b) Claim is inaccurate.

(c) Assuming the distribution is normally distributed:

$z = \dfrac{\bar{x} - \mu}{\sigma} = \dfrac{96.25 - 96}{0.5} = 0.5$

$P(x > 96.25) = P(z > 0.5) = 1 - P(z < 0.5) = 1 - 0.6915 = 0.3085$

Assuming the manufacturer's claim is true, an individual board with a length of 96.25 would not be unusual. It is within 1 standard deviation of the mean for an individual board.

37. (a) $\mu = 50{,}000$

$\sigma = 800$

$$z = \frac{\bar{x} - \mu}{\frac{\sigma}{\sqrt{n}}} = \frac{49{,}721 - 50{,}000}{\frac{800}{\sqrt{100}}} = \frac{-279}{80} = -3.49$$

$P(\bar{x} \le 49{,}721) = P(z \le -3.49) = 0.0002$

(b) The manufacturer's claim is inaccurate.

(c) Assuming the distribution is normally distributed:

$$z = \frac{x - \mu}{\sigma} = \frac{49{,}721 - 50{,}000}{800} = -0.35$$

$P(x < 49{,}721) = P(z < -0.35) = 0.3669$

Assuming the manufacturer's claim is true, an individual tire with a life span of 49,721 miles is not unusual. It is within 1 standard deviation of the mean for an individual tire.

39. $\mu = 518$

$\sigma = 115$

$$z = \frac{\bar{x} - \mu}{\frac{\sigma}{\sqrt{n}}} = \frac{530 - 518}{\frac{115}{\sqrt{50}}} = \frac{12}{16.26} = 0.74$$

$P(\bar{x} \ge 530) = P(z \ge 0.74) = 1 - P(z \le 0.74) = 1 - 0.7704 = 0.2296$

The high school's claim is not justified because it is not rare to find a sample mean as large as 530.

41. Use the finite correction factor since $n = 55 > 40 = 0.05N$.

$$z = \frac{\bar{x} - \mu}{\frac{\sigma}{\sqrt{n}}\sqrt{\frac{N-n}{N-1}}} = \frac{2.871 - 2.876}{\frac{0.009}{\sqrt{55}}\sqrt{\frac{800-55}{800-1}}} = \frac{-0.005}{(0.00121)\sqrt{0.9324}} \approx -4.27$$

$P(\bar{x} < 2.871) = P(z < -4.27) \approx 0$

43.

Sample	Number of boys	Proportion
bbb	3	1
bbg	2	$\frac{2}{3}$
bgb	2	$\frac{2}{3}$
bgg	1	$\frac{1}{3}$
gbb	2	$\frac{2}{3}$
gbg	1	$\frac{1}{3}$
ggb	1	$\frac{1}{3}$
ggg	0	0

45.

Sample	Sample Mean
bbb	1
bbg	$\frac{2}{3}$
bgb	$\frac{2}{3}$
bgg	$\frac{1}{3}$
gbb	$\frac{2}{3}$
gbg	$\frac{1}{3}$
ggb	$\frac{1}{3}$
ggg	0

The sample mean is the same as the proportion of boys in each sample.

47. $z = \dfrac{\hat{p} - p}{\sqrt{\dfrac{pq}{n}}} = \dfrac{0.70 - 0.75}{\sqrt{\dfrac{0.75(0.25)}{90}}} = \dfrac{-0.05}{0.0456} = -1.10$

$P(p < 0.70) = P(z < -1.10) = 0.1357$

5.5 NORMAL APPROXIMATIONS TO BINOMIAL DISTRIBUTIONS

5.5 Try It Yourself Solutions

1a. $n = 70, p = 0.80, q = 0.20$ **b.** $np = 56, nq = 14$

c. Because $np \geq 5$ and $nq \geq 5$, the normal distribution can be used.

d. $\mu = np = (70)(0.80) = 56$

$\sigma = \sqrt{npq} = \sqrt{(70)(0.80)(0.20)} \approx 3.35$

2a. (1) $57, 58, \ldots, 83$ (2) $\ldots, 52, 53, 54$

b. (1) $56.5 < x < 83.5$ (2) $x < 54.5$

3a. $n = 70, p = 0.80$

$np = 56 \geq 5$ and $nq = 14 \geq 5$

The normal distribution can be used.

b. $\mu = np = 56$

$\sigma = \sqrt{npq} \approx 3.35$

c. $x > 50.5$

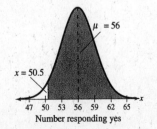

d. $z = \dfrac{x - \mu}{\sigma} = \dfrac{50.5 - 56}{3.35} \approx -1.64$

e. $P(z < -1.64) = 0.0505$

$P(x > 50.5) = P(z > -1.64) = 1 - P(z < -1.64) = 0.9495$

The probability that more than 50 respond yes is 0.9495.

4a. $n = 200, p = 0.38$

$np = 76 \geq 5$ and $nq = 124 \geq 5$

The normal distribution can be used.

b. $\mu = np = 76$

$\sigma = \sqrt{npq} \approx 6.86$

c. $P(x \leq 85.5)$

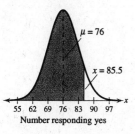

d. $z = \dfrac{x - \mu}{\sigma} = \dfrac{85.5 - 76}{6.86} \approx 1.38$

e. $P(x < 65.5) = P(z < 1.38) = 0.9162$

The probability that at most 65 people will say yes is 0.9162.

5a. $n = 200, p = 0.86$

$np = 172 \geq 5$ and $nq = 28 \geq 5$

The normal distribution can be used.

b. $\mu = np = 172$

$\sigma = \sqrt{npq} \approx 4.91$

c. $P(169.5 < x < 170.5)$

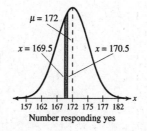

d. $z = \dfrac{x - \mu}{\sigma} = \dfrac{169.5 - 172}{4.91} \approx -0.51$

$z = \dfrac{x - \mu}{\sigma} = \dfrac{170.5 - 172}{4.91} \approx -0.31$

e. $P(z < -0.51) = 0.3050$

$P(z < -0.31) = 0.3783$

$P(-0.51 < z < -0.31) = 0.3783 - 0.3050 = 0.0733$

The probability that exactly 170 people will respond yes is 0.0733.

5.5 EXERCISE SOLUTIONS

1. $np = (24)(0.85) = 20.4 \geq 5$

$nq = (24)(0.15) = 3.6 < 5$

Cannot use normal distribution.

3. $np = (18)(0.90) = 16.2 \geq 5$

$nq = (18)(0.10) = 1.8 < 5$

Cannot use normal distribution.

5. $n = 10, p = 0.85, q = 0.15$

$np = 8.5 > 5, nq = 1.5 < 5$

Cannot use normal distribution because $nq < 5$.

7. $n = 10, p = 0.99, q = 0.03$

$np = 9.9 \geq 5, nq = 0.1 < 5$

Cannot use normal distribution
because $nq < 5$.

9. d   **11.** a   **13.** a   **15.** c

17. Binomial: $P(5 \leq x \leq 7) = 0.162 + 0.198 + 0.189 = 0.549$

Normal: $P(4.5 \leq x \leq 7.5) = P(-0.97 < z < 0.56) = 0.7123 - 0.1660 = 0.5463$

19. $n = 30, p = 0.07 \rightarrow np = 2.1$ and $nq = 27.9$

Cannot use normal distribution because $np < 5$.

(a) $P(x = 10) = {}_{30}C_{10}(0.07)^{10}(0.93)^{20} \approx 0.0000199$

(b) $P(x \geq 10) = 1 - P(x < 10)$

$$= 1 - [{}_{30}C_0(0.07)^0(0.93)^{30} + {}_{30}C_1(0.07)^1(0.93)^{29} + \cdots + {}_{30}C_9(0.07)^9(0.93)^{21}]$$

$$= 1 - .999977 \approx 0.000023$$

(c) $P(x < 10) \approx 0.999977$ (see part b)

(d) $n = 100, p = 0.07 \rightarrow np = 7$ and $nq = 93$

Use normal distribution.

$$z = \frac{x - \mu}{\sigma} = \frac{4.5 - 7}{2.55} \approx -0.98$$

$P(x < 5) = P(x < 4.5) = P(z < -0.98) = 0.1635$

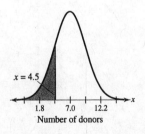

21. $n = 250, p = 0.05, q = 0.95$

$np = 12.5 \geq 5, nq = 237.5 \geq 5$

Use the normal distribution.

(a) $z = \dfrac{x - \mu}{\sigma} = \dfrac{15.5 - 12.5}{3.45} = 0.87$

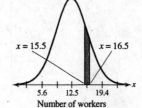

$z = \dfrac{x - \mu}{\sigma} = \dfrac{16.5 - 12.5}{3.45} = 1.16$

$P(x = 16) \approx P(15.5 \leq x \leq 16.5) = P(0.87 \leq z \leq 1.16)$

$= 0.8770 - 0.8078 = 0.0692$

(b) $P(x \geq 9) \approx P(x \geq 8.5) = P(z \geq -1.16) = 1 - P(z \leq -1.16) = 1 - 0.1230 = 0.8770$

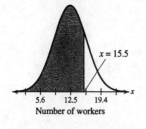

(c) $P(x < 16) \approx P(x \leq 15.5) = P(z \leq 0.87) = 0.8078$

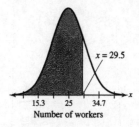

(d) $n = 500, p = 0.05, q = 0.95$

$np = 25 \geq 5, nq = 475 \geq 5$

Use normal distribution.

$z = \dfrac{x - \mu}{\sigma} = \dfrac{29.5 - 25}{4.87} = 0.92$

$P(x < 30) \approx P(x < 29.5) = P(z < 0.92) = 0.8212$

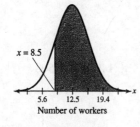

23. $n = 40, p = 0.52 \rightarrow np = 20.8$ and $nq = 19.2$

Use the normal distribution.

(a) $z = \dfrac{x - \mu}{\sigma} = \dfrac{23.5 - 20.8}{3.160} \approx 0.85$

$P(x \leq 23) = P(x < 23.5) = P(z < 0.85) = 0.8023$

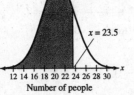

x = 23.5

12 14 16 18 20 22 24 26 28 30
Number of people

(b) $z = \dfrac{x - \mu}{\sigma} = \dfrac{17.5 - 20.8}{3.160} \approx -1.04$

$P(x \geq 18) = P(x > 17.5) = P(z > -1.04) = 1 - P(z < -1.04) = 1 - 0.1492 = 0.8508$

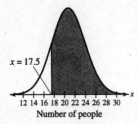

x = 17.5

12 14 16 18 20 22 24 26 28 30
Number of people

(c) $P(x > 20) = P(x > 20.5) = P(z > -0.09) = 1 - P(z < -0.09) = 1 - 0.4641 = 0.5359$

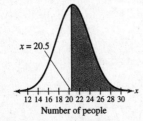

x = 20.5

12 14 16 18 20 22 24 26 28 30
Number of people

(d) $n = 650, p = 0.52 \rightarrow np = 338$ and $nq = 312$

Use normal distribution.

$z = \dfrac{x - \mu}{\sigma} = \dfrac{350.5 - 338}{12.74} \approx 0.98$

$P(x > 350) = P1(x > 350.5) = P(z > 0.98) = 1 - P(z < 0.98) = 1 - 0.8365 = 0.1635$

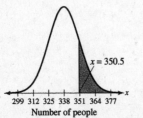

x = 350.5

299 312 325 338 351 364 377
Number of people

25. (a) $n = 25, p = 0.24, q = 0.76$

$np = 6 \geq 5, nq = 19 \geq 5$

Use normal distribution.

(b) $z = \dfrac{x - \mu}{\sigma} = \dfrac{8.5 - 6}{2.135} = 1.17$

$P(x > 8) \approx P(x \geq 8.5) = P(z \geq 1.17) = 1 - P(z \leq 1.17) = 1 - 0.8790 = 0.121$

(c) $z = \dfrac{x - \mu}{\sigma} = \dfrac{7.5 - 6}{2.135} = 0.70$

$z = \dfrac{x - \mu}{\sigma} = \dfrac{8.5 - 6}{2.135} = 1.17$

$P(x = 8) \approx P(7.5 \leq x \leq 8.5) = P(0.07 \leq z \leq 1.17) = 0.8790 - 0.7580 = 0.121$

It is not unusual for 8 out of 25 homeowners to say their home is too small because the z-score is within 1 standard deviation of the mean.

27. $n = 250, p = 0.70$

60% say no $\rightarrow$ 250(0.6) = 150 say no while 100 say yes.

$z = \dfrac{x - \mu}{\sigma} = \dfrac{99.5 - 175}{7.25} = -10.41$

$P(\text{less than 100 yes}) = P(x < 100) = P(x < 99.5) = P(z < -10.41) \approx 0$

It is highly unlikely that 60% responded no. Answers will vary.

29. $n = 100, p = 0.75$

$z = \dfrac{x - \mu}{\sigma} = \dfrac{69.5 - 75}{4.33} \approx 1.27$

$P(\text{reject claim}) = P(x < 70) = P(x < 69.5) = P(z < -1.27) = 0.1020$

CHAPTER 5 REVIEW EXERCISE SOLUTIONS

1. $\mu = 15, \sigma = 3$

3. $x = 1.32$: $z = \dfrac{x - \mu}{\sigma} = \dfrac{1.32 - 1.5}{0.08} = -2.25$

$x = 1.54$: $z = \dfrac{x - \mu}{\sigma} = \dfrac{1.54 - 1.5}{0.08} = 0.5$

$x = 1.66$: $z = \dfrac{x - \mu}{\sigma} = \dfrac{1.66 - 1.5}{0.08} = 2$

$x = 1.78$: $z = \dfrac{x - \mu}{\sigma} = \dfrac{1.78 - 1.5}{0.08} = 3.5$

5. 0.6293

7. 0.3936

9. $1 - 0.9535 = 0.0465$

11. $0.5 - 0.0505 = 0.4495$

13. $0.9564 - 0.5199 = 0.4365$

15. $0.0668 + 0.0668 = 0.1336$

17. $P(z < 1.28) = 0.8997$

19. $P(-2.15 < x < 1.55) = 0.9394 - 0.0158 = 0.9236$

21. $P(z < -2.50 \text{ or } z > 2.50) = 2(0.0062) = 0.0124$

23. (a) $z = \dfrac{x - \mu}{\sigma} = \dfrac{1900 - 2200}{625} \approx -0.48$

$P(x < 1900) = P(z < -0.48) = 0.3156$

(b) $z = \dfrac{x - \mu}{\sigma} = \dfrac{2000 - 2200}{625} \approx -0.32$

$z = \dfrac{x - \mu}{\sigma} = \dfrac{25,900 - 2200}{625} \approx 0.48$

$P(2000 < x < 2500) = P1(-0.32 < z < 0.48) = 0.6844 - 0.3745 = 0.3099$

(c) $z = \dfrac{x - \mu}{\sigma} = \dfrac{2450 - 2200}{625} = -0.4$

$P(x > 2450) = P1(z > 0.4) = 0.3446$

25. $z = -0.07$ **27.** $z = 1.13$ **29.** $z = 1.04$

31. $x = \mu + z\sigma = 52 + (-2.4)(2.5) = 46$ meters

33. 95th percentile $\Rightarrow$ Area $= 0.95 \Rightarrow z = 1.645$

$x = \mu + z\sigma = 52 + (1.645)(2.5) = 56.1$ meters

35. Top 10% $\Rightarrow$ Area $= 0.90 \Rightarrow z = 1.28$

$x = \mu + z\sigma = 52 + (1.28)(2.5) = 55.2$ meters

37. {90 90 90, 90 90 120, 90 90 160, 90 90 210, 90 90 300, 90 120 90, 90 120 120, 90 120 160, 90 120 210, 90 120 300, 90 160 90, 90 160 120, 90 160 160, 90 160 210, 90 160 300, 90 210 90, 90 210 120, 90 210 160, 90 210 210, 90 210 300, 90 300 90, 90 300 120, 90 300 160, 90 300 210, 90 300 300, 120 90 90, 120 90 120, 120 90 160, 120 90 210, 120 90 300, 120 120 90, 120 120 120, 120 120 160, 120 120 210, 120 120 300, 120 160 90, 120 160 120, 120 160 160, 120 160 210, 120 160 300, 120 210 90, 120 210 120, 120 210 160, 120 210 210, 120 210 300, 120 300 90, 120 300 120, 120 300 160, 120 300 210, 120 300 300, 160 90 90, 160 90 120, 160 90 160, 160 90 210, 160 90 300, 160 120 90, 160 120 120, 160 120 160, 160 120 210, 160 120 300, 160 160 90, 160 160 120, 160 160 160, 160 160 210, 160 160 300, 160 210 90, 160 210 120, 160 210 160, 160 210 210, 160 210 300, 160 300 90, 160 300 120, 160 300 160, 160 300 210, 160 300 300, 210 90 90, 210 90 120, 210 90 160, 210 90 210, 210 90 300, 210 120 90, 210 120 120, 210 120 160, 210 120 210, 210 120 300, 210 160 90, 210 160 120, 210 160 160, 210 160 210, 210 160 300, 210 210 90, 210 210 120, 210 210 160, 210 210 210, 210 210 300, 210 300 90, 210 300 120, 210 300 160, 210 300 210, 210 300 300, 300 90 90, 300 90 120, 300 90 160, 300 90 210, 300 90 300, 300 120 90, 300 120 120, 300 120 160, 300 120 210, 300 120 300, 300 160 90, 300 160 120, 300 160 160, 300 160 210, 300 160 300, 300 210 90, 300 210 120, 300 210 160, 300 210 210, 300 210 300, 300 300 90, 300 300 120, 300 300 160, 300 300 210, 300 300 300}

$\mu = 176, \sigma \approx 73.919$

$\mu_{\bar{x}} = 176, \sigma_{\bar{x}} \approx 42.677$

The means are the same, but $\sigma_{\bar{x}}$ is less than σ.

39. $\mu_{\bar{x}} = 144.3$, $\sigma_{\bar{x}} \dfrac{\sigma}{\sqrt{n}} = \dfrac{51.6}{\sqrt{35}} \approx 8.722$

126.9   144.3   161.7 $\bar{x}$

Mean consumption (in pounds)

41. (a) $z = \dfrac{\bar{x} - \mu}{\dfrac{\sigma}{\sqrt{n}}} = \dfrac{1900 - 2200}{\dfrac{625}{\sqrt{12}}} = \dfrac{-300}{180.42} \approx -1.66$

$P(\bar{x} < 1900) = P(z < -1.66) = 0.0485$

(b) $z = \dfrac{\bar{x} - \mu}{\dfrac{\sigma}{\sqrt{n}}} = \dfrac{2000 - 2200}{\dfrac{625}{\sqrt{12}}} = \dfrac{-200}{180.42} \approx -1.11$

$z = \dfrac{\bar{x} - \mu}{\dfrac{\sigma}{\sqrt{n}}} = \dfrac{2500 - 2200}{\dfrac{625}{\sqrt{12}}} = \dfrac{300}{180.42} \approx 1.66$

$P(2000 < \bar{x} < 2500) = P(-1.11 < z < 1.66) = 0.9515 - 0.1335 = 0.8180$

(c) $z = \dfrac{\bar{x} - \mu}{\dfrac{\sigma}{\sqrt{n}}} = \dfrac{2450 - 2200}{\dfrac{625}{\sqrt{12}}} = \dfrac{250}{180.42} \approx 1.39$

$P(\bar{x} > 2450) = P(z > 1.39) = 0.0823$

(a) and (c) are smaller, (b) is larger. This is to be expected because the standard error of the sample mean is smaller.

43. (a) $z = \dfrac{\bar{x} - \mu}{\dfrac{\sigma}{\sqrt{n}}} = \dfrac{29{,}000 - 29{,}200}{\dfrac{1500}{\sqrt{45}}} = \dfrac{-200}{223.61} \approx -0.89$

$P(\bar{x} < 29{,}000) = P(z < -0.89) \approx 0.1867$

(b) $z = \dfrac{\bar{x} - \mu}{\dfrac{\sigma}{\sqrt{n}}} = \dfrac{31{,}000 - 29{,}200}{\dfrac{1500}{\sqrt{45}}} = \dfrac{1500}{223.61} \approx 8.05$

$P(\bar{x} > 31{,}000) = P(z > 8.05) \approx 0$

45. Assuming the distribution is normally distributed:

$z = \dfrac{\bar{x} - \mu}{\dfrac{\sigma}{\sqrt{n}}} = \dfrac{1.125 - 1.5}{\dfrac{.5}{\sqrt{15}}} = \dfrac{-0.375}{0.129} \approx -2.90$

$P(\bar{x} < 1.125) = P(z < -2.90) = 0.0019$

47. $n = 12, p = 0.95, q = 0.05$

$np = 11.4 > 5$, but $nq = 0.6 < 5$

Cannot use the normal distribution because $nq < 5$.

49. $P(x \geq 25) = P(x > 24.5)$ **51.** $P(x = 45) = P(44.5 < x < 45.5)$

53. $n = 45, p = 0.70 \rightarrow np = 31.5, nq = 13.5$

Use normal distribution.

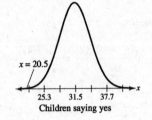

$\mu = np = 31.5, \sigma = \sqrt{npq} = \sqrt{45(0.70)(0.30)} \approx 3.07$

$z = \dfrac{x - \mu}{\sigma} = \dfrac{20.5 - 31.5}{3.07} \approx -3.58$

$P(x \leq 20) = P(x < 20.5) = P(z < -3.58) \approx 0$

CHAPTER 5 QUIZ SOLUTIONS

1. (a) $P(z > -2.10) = 0.9821$

(b) $P(z < 3.22) = 0.9994$

(c) $P(-2.33 < z < 2.33) = 0.9901 - 0.0099 = 0.9802$

(d) $P(z < -1.75 \text{ or } z > -0.75) = 0.0401 + 0.7734 = 0.8135$

2. (a) $z = \dfrac{x - \mu}{\sigma} = \dfrac{5.36 - 5.5}{0.08} \approx -1.75$

$z = \dfrac{x - \mu}{\sigma} = \dfrac{5.64 - 5.5}{0.08} \approx 1.75$

$P(5.36 < x < 5.64) = P(-1.75 < z < 1.75) = 0.9599 - 0.0401 = 0.9198$

(b) $z = \dfrac{x - \mu}{\sigma} = \dfrac{-5.00 - (-8.2)}{7.84} \approx 0.41$

$z = \dfrac{x - \mu}{\sigma} = \dfrac{0 - (-8.2)}{7.84} \approx 1.05$

$P(-5.00 < x < 0) = P(0.41 < z < 1.05) = 0.8531 - 0.6591 = 0.1940$

(c) $z = \dfrac{x - \mu}{\sigma} = \dfrac{0 - 18.5}{9.25} = -2$

$z = \dfrac{x - \mu}{\sigma} = \dfrac{37 - 18.5}{9.25} = 2$

$P(x < 0 \text{ or } x > 37) = P(z < -2 \text{ or } z > 2) = 2(0.0228) = 0.0456$

3. $z = \dfrac{x - \mu}{\sigma} = \dfrac{320 - 290}{37} \approx 0.81$

$P(x > 320) = P(z > 0.81) = 0.2090$

4. $z = \dfrac{x - \mu}{\sigma} = \dfrac{250 - 290}{37} \approx -1.08$

$z = \dfrac{x - \mu}{\sigma} = \dfrac{300 - 290}{37} \approx 0.27$

$P(250 < x < 300) = P(-1.08 < z < 0.27) = 0.6064 - 0.1401 = 0.4663$

5. $P(x > 250) = P(z > -1.08) = 0.8599 \rightarrow 85.99\%$

6. $z = \dfrac{x - \mu}{\sigma} = \dfrac{280 - 290}{37} \approx -0.27$

$P(x < 280) = P(z < -0.27) = 0.3936$

$(2000)(0.3936) = 787.2 \approx 787$ students

7. top $5\% \rightarrow z \approx 1.645$

$\mu + z\sigma = 290 + (1.645)(37) = 350.9 \approx 351$

8. bottom $25\% \rightarrow z \approx -0.67$

$\mu + z\sigma = 290 + (-0.67)(37) = 265.2 \approx 265$

9. $z = \dfrac{\bar{x} - \mu}{\dfrac{\sigma}{\sqrt{n}}} = \dfrac{300 - 290}{\dfrac{37}{\sqrt{60}}} = \dfrac{10}{4.78} \approx 2.09$

$P(\bar{x} > 300) = P(z > 2.09) \approx 0.0183$

10. $z = \dfrac{x - \mu}{\sigma} = \dfrac{300 - 290}{37} \approx 0.27$

$P(x > 300) = P(z > 0.27) = 0.3936$

$z = \dfrac{\bar{x} - \mu}{\dfrac{\sigma}{\sqrt{n}}} = \dfrac{300 - 290}{\dfrac{37}{\sqrt{15}}} = \dfrac{10}{4.78} \approx 2.09$

$P(\bar{x} > 300) = P(z > 2.09) = 0.0183$

You are more likely to select one student with a test score greater than 300 because the standard error of the mean is less than the standard deviation.

11. $n = 24, p = 0.75 \rightarrow np = 18, nq = 6$

Use normal distribution.

$\mu = np = 18 \qquad \sigma = \sqrt{npq} \approx 2.121$

12. $z = \dfrac{x - \mu}{\sigma} = \dfrac{15.5 - 18}{2.12} \approx -1.18$

$P(x \le 15) = P(x < 15.5) = P(z < -1.18) = 0.1190$

CUMULATIVE REVIEW, CHAPTERS 3–5

1. (a) $np = 30(0.56) = 16.8 > 5$

$np = 30(0.44) = 13.2 > 5$

Use normal distribution.

(b) $\mu = np = 30(0.56) = 16.8$

$\sigma = \sqrt{npq} = \sqrt{30(0.56)(0.44)} = 2.72$

$P(x \le 14) \approx P(x \le 14.5)$

$$= P\left(z \le \frac{14.5 - 16.8}{2.72}\right)$$

$$= P(z \le -0.85)$$

$$= 0.1977$$

(c) It is not unusual for 14 out of 30 employees to say they do not use all of their vacation time because the probability is greater than 0.05.

$P(x = 14) \approx P(13.5 \le x \le 14.5)$

$$= P\left(\frac{13.5 - 16.8}{2.72} \le z \le \frac{14.5 - 16.8}{2.72}\right)$$

$$= P(-1.21 \le z \le -0.85)$$

$$= 0.1977 - 0.1131$$

$$= 0.0846$$

2.

x	$P(x)$	$xP(x)$	$x - \mu$	$(x - \mu)^2$	$(x - \mu)^2 P(x)$
2	0.421	0.842	-1.131	1.279	0.539
3	0.233	0.699	-0.131	0.017	0.004
4	0.202	0.808	0.869	0.755	0.153
5	0.093	0.465	1.869	3.493	0.325
6	0.033	0.198	2.869	8.231	0.272
7	0.017	0.119	3.869	14.969	0.254
		$\sum xP(x) = 3.131$			$\sum (x - \mu)^2 P(x) = 1.546$

(a) $\mu = \sum xP(x) = 3.1$

(b) $\sigma^2 = \sum (x - \mu)^2 P(x) = 1.5$

(c) $\sigma = \sqrt{\sigma^2} = 1.2$

(d) $\sum(x) = \mu = 3.1$

(e) The expected family household size is 3.1 people with a standard deviation of 1.2 persons.

3.

x	$P(x)$	$xP(x)$	$x - \mu$	$(x - \mu)^2$	$(x - \mu)^2 P(x)$
0	0.012	0.000	-3.596	12.931	0.155
1	0.049	0.049	-2.596	6.739	0.330
2	0.159	0.318	-1.596	2.547	0.405
3	0.256	0.768	-0.596	0.355	0.091
4	0.244	0.976	0.404	0.163	0.040
5	0.195	0.975	1.404	1.971	0.384
6	0.085	0.510	2.404	5.779	0.491
		$\sum xP(x) = 3.596$			$\sum (x - \mu)^2 P(x) = 1.897$

(a) $\mu = \sum xP(x) \approx 3.6$

(b) $\sigma^2 = \sum (x - \mu)^2 P(x) \approx 1.9$

(c) $\sigma = \sqrt{\sigma^2} \approx 1.4$

(d) $\sum (x) = \mu \approx 3.6$

(e) The expected number of fouls per game is 3.6 with a standard deviation of 1.4 fouls.

4. (a) $P(x < 4) = 0.012 + 0.049 + 0.159 + 0.256 = 0.476$

(b) $P(x \geq 3) = 1 - P(x \leq 2)$

$$= 1 - (0.012 + 0.049 + 0.159)$$

$$= 0.78$$

(c) $P(2 \leq x \leq 4) = 0.159 + 0.256 + 0.244 = 0.659$

5. (a) $(16)(15)(14)(13) = 43{,}680$

(b) $\dfrac{(7)(6)(5)(4)}{(16)(15)(14)(13)} = \dfrac{840}{43{,}680} = 0.0192$

6. 0.9382   **7.** 0.0010

8. $1 - 0.2005 = 0.7995$ **9.** $0.9990 - 0.500 = 0.4990$

10. $0.3974 - 0.1112 = 0.2862$ **11.** $0.5478 + (1 - 0.9573) = 0.5905$

12. $n = 10, p = 0.78$

(a) $P(6) = 0.1108$

(b) $P(x \geq 6) = 0.9521$

(c) $P(x < 6)1 - P(x \geq 6) = 1 - 0.952 = 0.0479$

13. $p = \dfrac{1}{200} = 0.005$

(a) $P(x = 10) = (0.005)(0.995)^9 = 0.0048$

(b) $P(x \leq 3) = 0.0149$

(c) $P(x > 10) = 1 - P(x \leq 10) = 1 - 0.0489 = 0.9511$

14. (a) 0.224

(b) 0.879

(c) Dependent because a person can be a public school teacher and have more than 20 years experience.

(d) $0.879 + 0.137 - 0.107 = 0.909$

(e) $0.329 + 0.121 - 0.040 = 0.410$

15. (a) $\mu_x = 70$

$$\sigma_{\bar{x}} = \frac{\sigma}{\sqrt{n}} = \frac{1.2}{\sqrt{40}} = 0.1897$$

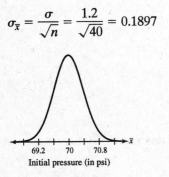

69.2   70   70.8 $\bar{x}$
Initial pressure (in psi)

(b) $P(\bar{x} \leq 69) = P\left(z \leq \dfrac{69 - 70}{\dfrac{1.2}{\sqrt{15}}}\right) = P(z < -3.23) = 0.0006$

16. (a) $P(x < 36) = P\left(z < \dfrac{36 - 44}{5}\right) = P(z < -1.6) = 0.0548$

(b) $P(42 < x < 60) = P\left(\dfrac{42 - 44}{5} < z < \dfrac{60 - 44}{5}\right)$

$$= P(-0.40 < z < 3.2)$$

$$= 0.9993 - 0.3446 = 0.6547$$

(c) Top $5\% \Rightarrow z = 1.645$

$$x = \mu + z\sigma = 44 + (1.645)(5) = 52.2 \text{ months}$$

17. (a) $_{12}C_4 = 495$

(b) $\dfrac{(1)(1)(1)(1)}{_{12}C_4} = 0.0020$

18. $n = 20, \ p = 0.41$

(a) $P(8) = 0.1790$

(b) $P(x \geq 6) = 1 - P(x \leq 5) = 1 - 0.108 = 0.8921$

(c) $P(x \leq 13) = 1 - P(x \geq 14) = 1 - 0.0084 = 0.9916$

Confidence Intervals

6.1 Try It Yourself Solutions

1a. $\bar{x} \approx 14.8$

b. The mean number of sentences per magazine advertisement is 14.8.

2a. $Z_c = 1.96, n = 30, s \approx 16.5$

b. $E = Z_c \dfrac{s}{\sqrt{n}} = 1.96 \dfrac{16.5}{\sqrt{30}} \approx 5.9$

c. You are 95% confident that the maximum error of the estimate is about 5.9 sentences per magazine advertisement.

3a. $\bar{x} \approx 14.8, E \approx 5.9$

b. $\bar{x} - E = 14.8 - 5.9 = 8.9$
$\bar{x} + E = 14.8 + 5.9 = 20.7$

c. You are 95% confident that the mean number of sentences per magazine advertisements is between 8.9 and 20.7.

4b. 75% CI: (11.6, 13.2)
85% CI: (11.4, 13.4)
99% CI: (10.6, 14.2)

c. The width of the interval increases as the level of confidence increases.

5a. $n = 30, \bar{x} = 22.9, \sigma = 1.5, Z_c = 1.645$

b. $E = Z_c \dfrac{\sigma}{\sqrt{n}} = 1.645 \dfrac{1.5}{\sqrt{30}} \approx 0.451 \approx 0.5$

$\bar{x} - E = 22.9 - 0.451 \approx 22.4$
$\bar{x} + E = 22.9 + 0.451 \approx 23.4$

c. You are 90% confident that the mean age of the students is between 22.4 and 23.4 years.

6a. $Z_c = 1.96, E = 2, s \approx 5.0$

b. $n = \left(\dfrac{Z_c s}{E}\right)^2 = \left(\dfrac{1.96 \cdot 5.0}{2}\right)^2 = 24.01 \rightarrow 25$

c. You should have at least 25 magazine advertisements in your sample.

6.1 EXERCISE SOLUTIONS

1. You are more likely to be correct using an interval estimate because it is unlikely that a point estimate will equal the population mean exactly.

3. d; As the level of confidence increases, z_c increases therefore creating wider intervals.

5. 1.28   **7.** 1.15

9. $\bar{x} - \mu = 3.8 - 4.27 = -0.47$

11. $\bar{x} - \mu = 26.43 - 24.67 = 1.76$

13. $\bar{x} - \mu = 0.7 - 1.3 = -0.60$

15. $E = z_c \dfrac{s}{\sqrt{n}} = 1.645 \dfrac{2.5}{\sqrt{36}} \approx 0.685$

17. $E = z_c \dfrac{s}{\sqrt{n}} = 0.93 \dfrac{1.5}{\sqrt{50}} = 0.197$

19. $c = 0.88 \Rightarrow z_c = 1.55$

$\bar{x} = 57.2, s = 7.1, n = 50$

$\bar{x} \pm z_c \dfrac{s}{\sqrt{n}} = 57.2 \pm 1.55 \dfrac{7.1}{\sqrt{50}} = 57.2 \pm 1.556 \approx (55.6, 58.8)$

Answer: (c)

21. $c = 0.95 \Rightarrow z_c = 1.96$

$\bar{x} = 57.2, s = 7.1, n = 50$

$\bar{x} \pm z_c \dfrac{s}{\sqrt{n}} = 57.2 \pm 1.96 \dfrac{7.1}{\sqrt{50}} = 57.2 \pm 1.968 \approx (55.2, 59.2)$

Answer: (b)

23. $\bar{x} \pm z_c \dfrac{s}{\sqrt{n}} = 15.2 \pm 1.645 \dfrac{2.0}{\sqrt{60}} = 15.2 \pm 0.425 \approx (14.8, 15.6)$

25. $\bar{x} \pm z_c \dfrac{s}{\sqrt{n}} = 4.27 \pm 1.96 \dfrac{0.3}{\sqrt{42}} = 4.27 \pm 0.091 \approx (4.18, 4.36)$

27. $(0.264, 0.494) \Rightarrow 0.379 \pm 0.115 \Rightarrow \bar{x} = 0.379, E = 0.115$

29. $(1.71, 2.05) \Rightarrow 1.88 \pm 0.17 \Rightarrow \bar{x} = 1.88, E = 0.17$

31. $c = 0.90 \Rightarrow z_c = 1.645$

$n = \left(\dfrac{z_c \sigma}{E}\right)^2 = \left(\dfrac{(1.645)(6.8)}{1}\right)^2 = 125.13 \Rightarrow 126$

33. $c = 0.80 \Rightarrow z_c = 1.28$

$n = \left(\dfrac{z_c \sigma}{E}\right)^2 = \left(\dfrac{(1.28)(4.1)}{2}\right)^2 = 6.89 \Rightarrow 7$

35. $(2.1, 3.5) \Rightarrow 2E = 3.5 - 2.1 = 1.4 \Rightarrow E = 0.7$ and $\bar{x} = 2.1 + E$
$$= 2.1 + 0.7 = 2.8$$

37. 90% CI: $\bar{x} \pm z_c \dfrac{s}{\sqrt{n}} = 630.90 \pm 1.645 \dfrac{56.70}{\sqrt{32}} = 630.9 \pm 16.49 \approx (614.41, 647.39)$

95% CI: $\bar{x} \pm z_c \dfrac{s}{\sqrt{n}} = 630.90 \pm 1.96 \dfrac{56.70}{\sqrt{32}} = 630.9 \pm 19.65 \approx (611.25, 650.55)$

The 95% confidence interval is wider.

39. 90% CI: $\bar{x} \pm z_c \dfrac{s}{\sqrt{n}} = 99.3 \pm 1.645 \dfrac{41.5}{\sqrt{31}} = 99.3 \pm 12.26 \approx (87.0, 111.6)$

   95% CI: $\bar{x} \pm z_c \dfrac{s}{\sqrt{n}} = 99.3 \pm 1.96 \dfrac{41.5}{\sqrt{31}} = 99.3 \pm 14.61 \approx (84.7, 113.9)$

   The 95% confidence interval is wider.

41. $\bar{x} \pm z_c \dfrac{s}{\sqrt{n}} = 120 \pm 1.96 \dfrac{17.50}{\sqrt{40}} = 120 \pm 5.423 \approx (114.58, 125.42)$

43. $\bar{x} \pm z_c \dfrac{s}{\sqrt{n}} = 120 \pm 1.96 \dfrac{17.50}{\sqrt{80}} = 120 \pm 3.8348 \approx (116.17, 123.83)$

   $n = 40$ CI is wider because a smaller sample was taken giving less information about the population.

45. $\bar{x} \pm z_c \dfrac{s}{\sqrt{n}} = 3.12 \pm 2.575 \dfrac{0.09}{\sqrt{48}} = 3.12 \pm 0.033 \approx (3.09, 3.15)$

47. $\bar{x} \pm z_c \dfrac{s}{\sqrt{n}} = 3.12 \pm 2.575 \dfrac{0.06}{\sqrt{48}} = 3.12 \pm 0.022 \approx (3.10, 3.14)$

   $s = 0.09$ CI is wider because of the increased variability within the sample.

49. (a) An increase in the level of confidence will widen the confidence interval.

   (b) An increase in the sample size will narrow the confidence interval.

   (c) An increase in the standard deviation will widen the confidence interval.

51. $\bar{x} = \dfrac{\Sigma x}{n} = \dfrac{136}{15} \approx 9.1$

   90% CI: $\bar{x} \pm z_c \dfrac{\sigma}{\sqrt{n}} = 9.1 \pm 1.645 \dfrac{1.5}{\sqrt{15}} = 9.1 \pm 0.637 \approx (8.4, 9.7)$

   99% CI: $\bar{x} \pm z_c \dfrac{\sigma}{\sqrt{n}} = 9.1 \pm 2.575 \dfrac{1.5}{\sqrt{15}} = 9.1 \pm 0.997 \approx (8.1, 10.1)$

   99% CI is wider.

53. $n = \left(\dfrac{z_c \sigma}{E}\right)^2 = \left(\dfrac{1.96 \cdot 4.8}{1}\right)^2 \approx 88.510 \rightarrow 89$

55. (a) $n = \left(\dfrac{z_c \sigma}{E}\right)^2 = \left(\dfrac{1.96 \cdot 2.8}{0.5}\right)^2 \approx 120.473 \rightarrow 121$ servings

   (b) $n = \left(\dfrac{z_c \sigma}{E}\right)^2 = \left(\dfrac{2.575 \cdot 2.8}{0.5}\right)^2 \approx 207.936 \rightarrow 208$ servings

   99% CI requires larger sample because more information is needed from the population to be 99% confident.

57. (a) $n = \left(\dfrac{z_c\sigma}{E}\right)^2 = \left(\dfrac{1.645 \cdot 0.85}{0.25}\right)^2 \approx 31.282 \rightarrow 32$ cans

(b) $n = \left(\dfrac{z_c\sigma}{E}\right)^2 = \left(\dfrac{1.645 \cdot 0.85}{0.15}\right)^2 \approx 86.893 \rightarrow 87$ cans

$E = 0.15$ requires a larger sample size. As the error size decreases, a larger sample must be taken to obtain enough information from the population to ensure desired accuracy.

59. $n = \left(\dfrac{z_c\sigma}{E}\right)^2 = \left(\dfrac{1.960 \cdot 0.25}{0.125}\right)^2 = 15.3664 \rightarrow 16$ sheets

$n = \left(\dfrac{z_c\sigma}{E}\right)^2 = \left(\dfrac{1.960 \cdot 0.25}{0.0625}\right)^2 = 61.4656 \rightarrow 62$ sheets

$E = 0.0625$ requires a larger sample size. As the error size decreases, a larger sample must be taken to obtain enough information from the population to ensure desired accuracy.

61. (a) $n = \left(\dfrac{z_c\sigma}{E}\right)^2 = \left(\dfrac{2.575 \cdot 0.25}{0.1}\right)^2 \approx 41.441 \rightarrow 42$ soccer balls

(b) $n = \left(\dfrac{z_c\sigma}{E}\right)^2 = \left(\dfrac{2.575 \cdot 0.30}{0.1}\right)^2 \approx 59.676 \rightarrow 60$ soccer balls

$\sigma = 0.30$ requires a larger sample size. Due to the increased variability in the population, a larger sample size is needed to ensure the desired accuracy.

63. (a) An increase in the level of confidence will increase the minimum sample size required.

(b) An increase (larger E) in the error tolerance will decrease the minimum sample size required.

(c) An increase in the population standard deviation will increase the minimum sample size required.

65. $\bar{x} = 238.77, s \approx 13.20, n = 31$

$\bar{x} \pm z_c\dfrac{s}{\sqrt{n}} = 238.77 \pm 1.96\dfrac{13.20}{\sqrt{31}} = 238.77 \pm 4.65 \approx (234.1, 243.4)$

67. $\bar{x} = 15.783, s \approx 2.464, n = 30$

$\bar{x} \pm z_c\dfrac{s}{\sqrt{n}} = 15.783 \pm 1.96\dfrac{2.464}{\sqrt{30}} = 15.783 \pm 0.88173 \approx (14.902, 16.665)$

69. (a) $\sqrt{\dfrac{N-n}{N-1}} = \sqrt{\dfrac{1000-500}{1000-1}} \approx 0.707$

(b) $\sqrt{\dfrac{N-n}{N-1}} = \sqrt{\dfrac{1000-100}{1000-1}} \approx 0.949$

(c) $\sqrt{\dfrac{N-n}{N-1}} = \sqrt{\dfrac{1000-75}{1000-1}} \approx 0.962$

(d) $\sqrt{\dfrac{N-n}{N-1}} = \sqrt{\dfrac{1000-50}{1000-1}} \approx 0.975$

(e) The finite population correction factor approaches 1 as the sample size decreases while the population size remains the same.

71. $E = \dfrac{z_c\sigma}{\sqrt{n}} \rightarrow \sqrt{n} = \dfrac{z_c\sigma}{E} \rightarrow n = \left(\dfrac{z_c\sigma}{E}\right)^2$

6.2 CONFIDENCE INTERVALS FOR THE MEAN (SMALL SAMPLES)

6.2 Try It Yourself Solutions

1a. d.f. $= n - 1 = 22 - 1 = 21$

b. $c = 0.90$

c. 1.721

2a. 90% CI: $t_c = 1.753$

$$E = t_c \frac{s}{\sqrt{n}} = 1.753 \frac{10}{\sqrt{16}} \approx 4.383 \approx 4.4$$

99% CI: $t_c = 2.947$

$$E = t_c \frac{s}{\sqrt{n}} = 2.947 \frac{10}{\sqrt{16}} \approx 7.368 \approx 7.4$$

b. 90% CI: $\bar{x} \pm E = 162 \pm 4.383 \approx (157.6, 166.4)$
   99% CI: $\bar{x} \pm E = 162 \pm 7.368 \approx (154.6, 169.4)$

c. You are 90% confident that the mean temperature of coffee sold is between 157.6°
   and 166.4°.
   You are 99% confident that the mean temperature of coffee sold is between 154.6°
   and 169.4°.

3a. 90% CI: $t_c = 1.729$

$$E = t_c \frac{s}{\sqrt{n}} = 1.729 \frac{0.42}{\sqrt{20}} \approx 0.162 \approx 0.16$$

95% CI: $t_c = 2.093$

$$E = t_c \frac{s}{\sqrt{n}} = 2.093 \frac{0.42}{\sqrt{20}} \approx 0.197 \approx 0.20$$

b. 90% CI: $\bar{x} \pm E = 6.22 \pm 0.162 \approx (6.06, 6.38)$
   95% CI: $\bar{x} \pm E = 6.22 \pm 0.197 \approx (6.02, 6.42)$

c. You are 90% confident that the mean mortgage interest rate is contained between 6.06%
   and 6.38%.
   You are 95% confident that the mean mortgage interest rate is contained between 6.02%
   and 6.42%.

4a. Is $n \geq 30$? No
   Is the population normally distributed? Yes
   Is σ known? No
   Use the t-distribution to construct the 90% CI.

6.2 EXERCISE SOLUTIONS

1. 1.833   **3.** 2.947

5. $E = t_c \frac{s}{\sqrt{n}} = 2.131 \frac{5}{\sqrt{16}} \approx 2.664 \approx 2.7$

7. $E = t_c \dfrac{s}{\sqrt{n}} = 1.796 \dfrac{2.4}{\sqrt{12}} = 1.244 \approx 1.2$

9. (a) $\bar{x} \pm t_c \dfrac{s}{\sqrt{n}} = 12.5 \pm 2.015 \dfrac{2.0}{\sqrt{6}} = 12.5 \pm 1.645 \approx (10.9, 14.1)$

(b) $\bar{x} \pm t_c \dfrac{s}{\sqrt{n}} = 12.5 \pm 1.645 \dfrac{2.0}{\sqrt{6}} = 12.5 \pm 1.343 \approx (11.2, 13.8)$

t-CI is wider.

11. (a) $\bar{x} \pm t_c \dfrac{s}{\sqrt{n}} = 4.3 \pm 2.650 \dfrac{0.34}{\sqrt{14}} = 4.3 \pm 0.241 \approx (4.1, 4.5)$

(b) $\bar{x} \pm z_c \dfrac{s}{\sqrt{n}} = 4.3 \pm 2.326 \dfrac{0.34}{\sqrt{14}} = 4.3 \pm 0.211 \approx (4.1, 4.5)$

Both CIs have the same width.

13. $\bar{x} \pm t_c \dfrac{s}{\sqrt{n}} = 75 \pm 2.776 \dfrac{12.50}{\sqrt{5}} = 75 \pm 15.518 \approx (59.48, 90.52)$

$E = t_c \dfrac{s}{\sqrt{n}} = 2.776 \dfrac{12.50}{\sqrt{5}} \approx 15.518$

15. $\bar{x} \pm z_c \dfrac{\sigma}{\sqrt{n}} = 75 \pm 1.96 \dfrac{15}{\sqrt{5}} = 75 \pm 13.148 \approx (61.85, 88.15)$

$E = z_c \dfrac{\sigma}{\sqrt{n}} = 1.96 \dfrac{15}{\sqrt{5}} \approx 13.148 \approx 13.15$

t-CI is wider.

17. (a) $\bar{x} \pm t_c \dfrac{s}{\sqrt{n}} = 4.54 \pm 1.833 \dfrac{1.21}{\sqrt{10}} = 4.54 \pm 0.701 \approx (3.84, 5.24)$

(b) $\bar{x} \pm z_c \dfrac{s}{\sqrt{n}} = 4.54 \pm 1.645 \dfrac{1.21}{\sqrt{500}} = 4.54 \pm 0.089 \approx (4.45, 4.63)$

t-CI is wider.

19. (a) $\bar{x} = 4460.16$

(b) $s \approx 146.143$

(c) $\bar{x} \pm t_c \dfrac{s}{\sqrt{n}} = 4460.16 \pm 3.250 \dfrac{146.143}{\sqrt{10}} = 4460.16 \pm 150.197 \approx (4309.96, 4610.36)$

21. (a) $\bar{x} \approx 1767.7$

(b) $s \approx 252.23 \approx 252.2$

(c) $\bar{x} \pm t_c \dfrac{s}{\sqrt{n}} = 1767.7 \pm 3.106 \dfrac{252.23}{\sqrt{12}} = 1767.7 \pm 226.16 \approx (1541.5, 1993.8)$

23. $n \geq 30 \rightarrow$ use normal distribution

$\bar{x} \pm z_c \dfrac{s}{\sqrt{n}} = 1.25 \pm 1.96 \dfrac{0.05}{\sqrt{70}} = 1.25 \pm 0.012 \approx (1.24, 1.26)$

25. $\bar{x} = 21.9$, $s = 3.46$, $n < 30$, σ known, and pop normally distributed → use t-distribution

$$\bar{x} \pm t_c\frac{s}{\sqrt{n}} = 21.9 \pm 2.064\frac{3.46}{\sqrt{25}} = 21.9 \pm 1.43 \approx (20.5, 23.3)$$

27. $n < 30$, σ unknown, and pop *not* normally distributed → cannot use either the normal or t-distributions.

29. $n = 25$, $\bar{x} = 56.0$, $s = 0.25$

$\pm t_{0.99} \rightarrow$ 99% t-CI

$$\bar{x} \pm t_c\frac{s}{\sqrt{n}} = 56.0 \pm 2.797\frac{0.25}{\sqrt{25}} = 56.0 \pm 0.140 \approx (55.9, 56.1)$$

They are not making good tennis balls because desired bounce height of 55.5 inches is not contained between 55.9 and 56.1 inches.

6.3 CONFIDENCE INTERVALS FOR POPULATION PROPORTIONS

6.3 Try It Yourself Solutions

1a. $x = 181$, $n = 1006$

b. $\hat{p} = \dfrac{181}{1006} \approx 0.180$

2a. $\hat{p} \approx 0.180$, $\hat{q} \approx 0.820$

b. $n\hat{p} = (1006)(0.180) = 181.08 > 5$

$n\hat{q} = (1006)(0.820) = 824.92 > 5$

c. $z_c = 1.645$

$$E = z_c\sqrt{\frac{\hat{p}\hat{q}}{n}} = 1.645\sqrt{\frac{0.180 \cdot 0.820}{1006}} \approx 0.020$$

d. $\hat{p} \pm E = 0.180 \pm 0.020 \approx (0.160, 0.200)$

e. You are 90% confident that the proportion of adults say that Abraham Lincoln was the greatest president is contained between 16.0% and 20.0%.

3a. $n = 900$, $\hat{p} \approx 0.33$

b. $\hat{q} = 1 - \hat{p} = 1 - 0.33 \approx 0.67$

c. $n\hat{p} = 900 \cdot 0.33 \approx 297 > 5$

$n\hat{q} = 900 \cdot 0.67 \approx 603 > 5$

Distribution of $\hat{p}$ is approximately normal.

d. $z_c = 2.575$

e. $\hat{p} \pm z_c\sqrt{\dfrac{\hat{p}\hat{q}}{n}} = 0.33 \pm 2.575\sqrt{\dfrac{0.33 \cdot 0.67}{900}} = 0.33 \pm 0.04 \approx (0.290, 0.370)$

f. You are 99% confident that the proportion of adults who think that people over 75 are more dangerous drivers is contained between 29.0% and 37.0%.

4a. (1) $\hat{p} = 0.50$, $\hat{q} = 0.50$

$z_c = 1.645$, $E = 0.02$

(2) $\hat{p} = 0.064$, $\hat{q} = 0.936$

$z_c = 1.645$, $E = 0.02$

b. (1) $n = \hat{p}\hat{q}\left(\dfrac{z_c}{E}\right)^2 = (0.50)(0.50)\left(\dfrac{1.645}{0.02}\right)^2 = 1691.266 \rightarrow 1692$

(2) $n = \hat{p}\hat{q}\left(\dfrac{z_c}{E}\right)^2 = 0.064 \cdot 0.936\left(\dfrac{1.645}{0.02}\right)^2 \approx 405.25 \rightarrow 406$

c. (1) At least 1692 males should be included in the sample.

(2) At least 406 males should be included in the sample.

6.3 EXERCISE SOLUTIONS

1. False. To estimate the value of p, the population proportion of successes, use the point estimate $\hat{p} = \dfrac{x}{n}$.

3. $\hat{p} = \dfrac{x}{n} = \dfrac{752}{1002} \approx 0.750$

$\hat{q} = 1 - \hat{p} \approx 0.250$

5. $\hat{p} = \dfrac{x}{n} = \dfrac{2938}{4431} \approx 0.663$

$\hat{q} = 1 - \hat{p} \approx 0.337$

7. $\hat{p} = \dfrac{x}{n} = \dfrac{144}{848} \approx 0.170$

$\hat{q} = 1 - \hat{p} \approx 0.830$

9. $\hat{p} = \dfrac{x}{n} = \dfrac{204}{284} \approx 0.718$

$\hat{q} = 1 - \hat{p} \approx 0.282$

11. $\hat{p} = \dfrac{x}{n} = \dfrac{230}{1000} = 0.230$

$q = 1 - \hat{p} = 1 - 0.230 = 0.770$

13. $\hat{p} = 0.48$, $E = 0.03$

$\hat{p} \pm E = 0.48 \pm 0.03 = (0.45, 0.51)$

Note: Exercises 15–19 may have slightly different answers from the text due to rounding.

15. 95% CI: $\hat{p} \pm z_c\sqrt{\dfrac{\hat{p}\hat{q}}{n}} = 0.750 \pm 1.96\sqrt{\dfrac{0.750 \cdot 0.25}{1002}} = 0.750 \pm 0.027 \approx (0.723, 0.777)$

99% CI: $\hat{p} \pm z_c\sqrt{\dfrac{\hat{p}\hat{q}}{n}} = 0.750 \pm 2.575\sqrt{\dfrac{0.750 \cdot 0.250}{1002}} = 0.75 \pm 0.035 \approx (0.715, 0.785)$

99% CI is wider.

17. 95% CI: $\hat{p} \pm z_c\sqrt{\dfrac{\hat{p}\hat{q}}{n}} = 0.663 \pm 1.96\sqrt{\dfrac{0.663 \cdot 0.337}{4431}} = 0.663 \pm 0.014 \approx (0.649, 0.677)$

99% CI: $\hat{p} \pm z_c\sqrt{\dfrac{\hat{p}\hat{q}}{n}} = 0.663 \pm 2.575\sqrt{\dfrac{0.663 \cdot 0.337}{4431}} = 0.663 \pm 0.018 \approx (0.645, 0.681)$

99% CI is wider.

19. 95% CI: $\hat{p} \pm z_c \sqrt{\dfrac{\hat{p}\hat{q}}{n}} = 0.170 \pm 1.96 \sqrt{\dfrac{0.170 \cdot 0.830}{848}} = 0.170 \pm 0.025 \approx (0.145, 0.195)$

    99% CI: $\hat{p} \pm z_c \sqrt{\dfrac{\hat{p}\hat{q}}{n}} = 0.170 \pm 2.575 \sqrt{\dfrac{0.170 \cdot 0.830}{848}} = 0.170 \pm 0.033 \approx (0.137, 0.203)$

    99% CI is wider.

21. (a) $n = \hat{p}\hat{q}\left(\dfrac{z_c}{E}\right)^2 = 0.5 \cdot 0.5 \left(\dfrac{1.96}{0.03}\right)^2 \approx 1067.111 \rightarrow 1068$ vacationers

    (b) $n = \hat{p}\hat{q}\left(\dfrac{z_c}{E}\right)^2 = 0.26 \cdot 0.74 \left(\dfrac{1.96}{0.03}\right)^2 \approx 821.249 \rightarrow 822$ vacationers

    (c) Having an estimate of the proportion reduces the minimum sample size needed.

23. (a) $n = \hat{p}\hat{q}\left(\dfrac{z_c}{E}\right)^2 = 0.5 \cdot 0.5 \left(\dfrac{2.05}{0.025}\right)^2 = 1681$ camcorders

    (b) $n = \hat{p}\hat{q}\left(\dfrac{z_c}{E}\right)^2 = 0.25 \cdot 0.75 \left(\dfrac{2.05}{0.025}\right)^2 \approx 1260.75 \rightarrow 1261$ camcorders

    (c) Having an estimate of the proportion reduces the minimum sample size needed.

25. (a) $\hat{p} = 0.36, n = 400$

    $\hat{p} \pm z_c \sqrt{\dfrac{\hat{p}\hat{q}}{n}} = 0.36 \pm 2.575 \sqrt{\dfrac{0.36 \cdot 0.64}{400}} = 0.36 \pm 0.062 \approx (0.298, 0.422)$

    (b) $\hat{p} = 0.32, n = 400$

    $\hat{p} \pm z_c \sqrt{\dfrac{\hat{p}\hat{q}}{n}} = 0.32 \pm 2.575 \sqrt{\dfrac{0.32 \cdot 0.68}{400}} = 0.32 \pm 0.060 \approx (0.260, 0.380)$

    It is possible that the two proportions are equal because the confidence intervals estimating the proportions overlap.

27. (a) $\hat{p} = 0.65, \hat{q} = 0.35, n = 2563$

    $\hat{p} \pm z_c \sqrt{\dfrac{\hat{p}\hat{q}}{n}} = 0.65 \pm 2.575 \sqrt{\dfrac{(0.65)(0.35)}{2563}} = 0.65 \pm 0.024 = (0.626, 0.674)$

    (b) $\hat{p} = 0.88, \hat{q} = 0.12, n = 1125$

    $\hat{p} \pm z_c \sqrt{\dfrac{\hat{p}\hat{q}}{n}} = 0.88 \pm 2.575 \sqrt{\dfrac{(0.88)(0.12)}{1125}} = 0.88 \pm 0.025 = (0.855, 0.905)$

    (c) $\hat{p} = 0.92, \hat{q} = 0.08, n = 1086$

    $\hat{p} \pm z_c \sqrt{\dfrac{\hat{p}\hat{q}}{n}} = 0.92 \pm 2.575 \sqrt{\dfrac{(0.92)(0.08)}{1086}} = 0.92 \pm 0.021 = (0.899, 0.941)$

29. $31.4\% \pm 1\% \rightarrow (30.4\%, 32.4\%) \rightarrow (0.304, 0.324)$

    $E = z_c \sqrt{\dfrac{\hat{p}\hat{q}}{n}} \rightarrow z_c = E\sqrt{\dfrac{n}{\hat{p}\hat{q}}} = 0.01 \sqrt{\dfrac{8451}{0.314 \cdot 0.686}} \approx 1.981 \rightarrow z_c = 1.98 \rightarrow c = 0.952$

    $(30.4\%, 32.4\%)$ is approximately a 95.2% CI.

31. If $n\hat{p} < 5$ or $n\hat{q} < 5$, the sampling distribution of $\hat{p}$ may not be normally distributed, therefore preventing the use of z_c when calculating the confidence interval.

33.

$\hat{p}$	$\hat{q} = 1 - \hat{p}$	$\hat{p}\hat{q}$	$\hat{p}$	$\hat{q} = 1 - \hat{p}$	$\hat{p}\hat{q}$
0.0	1.0	0.00	0.45	0.55	0.2475
0.1	0.9	0.09	0.46	0.54	0.2484
0.2	0.8	0.16	0.47	0.53	0.2491
0.3	0.7	0.21	0.48	0.52	0.2496
0.4	0.6	0.24	0.49	0.51	0.2499
0.5	0.5	0.25	0.50	0.50	0.2500
0.6	0.4	0.24	0.51	0.49	0.2499
0.7	0.3	0.21	0.52	0.48	0.2496
0.8	0.2	0.16	0.53	0.47	0.2491
0.9	0.1	0.09	0.54	0.46	0.2484
1.0	0.0	0.00	0.55	0.45	0.2475

$\hat{p} = 0.5$ give the maximum value of $\hat{p}\hat{q}$.

6.4 CONFIDENCE INTERVALS FOR VARIANCE AND STANDARD DEVIATION

6.4 Try It Yourself Solutions

1a. d.f. $= n - 1 = 24$

level of confidence $= 0.95$

b. Area to the right of χ_R^2 is 0.025.

Area to the left of χ_L^2 is 0.975.

c. $\chi_R^2 = 39.364$, $\chi_L^2 = 12.401$

2a. 90% CI: $\chi_R^2 = 42.557$, $\chi_L^2 = 17.708$

95% CI: $\chi_R^2 = 45.722$, $\chi_L^2 = 16.047$

b. 90% CI for σ^2: $\left(\dfrac{(n-1)s^2}{\chi_R^2}, \dfrac{(n-1)s^2}{\chi_L^2} \right) = \left(\dfrac{29 \cdot (1.2)^2}{42.557}, \dfrac{29 \cdot (1.2)^2}{17.708} \right) \approx (0.98, 2.36)$

95% CI for σ^2: $\left(\dfrac{(n-1)s^2}{\chi_R^2}, \dfrac{(n-1)s^2}{\chi_L^2} \right) = \left(\dfrac{29 \cdot (1.2)^2}{45.722}, \dfrac{29 \cdot (1.2)^2}{16.047} \right) \approx (0.91, 2.60)$

c. 90% CI for σ: $\left(\sqrt{0.981}, \sqrt{2.358} \right) = (0.99, 1.54)$

95% CI for σ: $\left(\sqrt{0.913}, \sqrt{2.602} \right) = (0.96, 1.61)$

d. You are 90% confident that the population variance is between 0.98 and 2.36, and that the population standard deviation is between 0.99 and 1.54. You are 95% confident that the population variance is between 0.91 and 2.60, and that the population standard deviation is between 0.96 and 1.61.

6.4 EXERCISE SOLUTIONS

1. $\chi_R^2 = 16.919$, $\chi_L^2 = 3.325$ **3.** $\chi_R^2 = 35.479$, $\chi_L^2 = 10.283$ **5.** $\chi_R^2 = 52.336$, $\chi_L^2 = 13.121$

7. (a) $s = 0.00843$

$$\left(\frac{(n-1)s^2}{\chi_R^2}, \frac{(n-1)s^2}{\chi_L^2}\right) = \left(\frac{13 \cdot (0.00843)^2}{22.362}, \frac{13 \cdot (0.00843)^2}{5.892}\right) \approx (0.0000413, 0.000157)$$

(b) $\left(\sqrt{0.0000413}, \sqrt{0.000157}\right) \approx (0.00643, 0.0125)$

9. (a) $s = 0.253$

$$\left(\frac{(n-1)s^2}{\chi_R^2}, \frac{(n-1)s^2}{\chi_L^2}\right) = \left(\frac{17 \cdot (0.253)^2}{35.718}, \frac{17 \cdot (0.253)^2}{5.697}\right) \approx (0.0305, 0.191)$$

(b) $\left(\sqrt{0.0305}, \sqrt{0.191}\right) \approx (0.175, 0.437)$

11. (a) $\left(\frac{(n-1)s^2}{\chi_R^2}, \frac{(n-1)s^2}{\chi_L^2}\right) = \left(\frac{11 \cdot (3.25)^2}{26.757}, \frac{11 \cdot (3.25)^2}{2.603}\right) \approx (4.34, 44.64)$

(b) $\left(\sqrt{4.342}, \sqrt{44.636}\right) \approx (2.08, 6.68)$

13. (a) $\left(\frac{(n-1)s^2}{\chi_R^2}, \frac{(n-1)s^2}{\chi_L^2}\right) = \left(\frac{9 \cdot (26)^2}{16.919}, \frac{9 \cdot (26)^2}{3.325}\right) \approx (359.6, 1829.8)$

(b) $\left(\sqrt{359.596}, \sqrt{1829.774}\right) \approx (19.0, 42.8)$

15. (a) $\left(\frac{(n-1)s^2}{\chi_R^2}, \frac{(n-1)s^2}{\chi_L^2}\right) = \left(\frac{18 \cdot (15)^2}{31.526}, \frac{18 \cdot (15)^2}{8.231}\right) \approx (128, 492)$

(b) $\left(\sqrt{128.465}, \sqrt{492.042}\right) \approx (11, 22)$

17. (a) $\left(\frac{(n-1)s^2}{\chi_R^2}, \frac{(n-1)s^2}{\chi_L^2}\right) = \left(\frac{13 \cdot (342)^2}{24.736}, \frac{13 \cdot (342)^2}{5.009}\right) \approx (61,470, 303,559)$

(b) $\left(\sqrt{61,470.41}, \sqrt{303,559.99}\right) \approx (248, 551)$

19. (a) $\left(\frac{(n-1)s^2}{\chi_R^2}, \frac{(n-1)s^2}{\chi_L^2}\right) = \left(\frac{(21)(3.6)^2}{38.932}, \frac{(21)(3.6)^2}{8.897}\right) \approx (7.0, 30.6)$

(b) $\left(\sqrt{6.99}, \sqrt{30.59}\right) = (2.6, 5.5)$

21. 90% CI for σ: $(0.00643, 0.0125)$

Yes, because the confidence interval is below 0.015.

CHAPTER 6 REVIEW EXERCISE SOLUTIONS

1. (a) $\bar{x} \approx 103.5$

(b) $s \approx 34.663$

$$E = z_c \frac{s}{\sqrt{n}} = 1.645 \frac{34.663}{\sqrt{40}} \approx 9.0$$

3. $\bar{x} \pm z_c \frac{s}{\sqrt{n}} = 10.3 \pm 1.96 \frac{0.277}{\sqrt{100}} = 10.3 \pm 0.054 \approx (10.2, 10.4)$

5. $s = 34.663$

$$n = \left(\frac{z_c \sigma}{E}\right)^2 = \left(\frac{1.96 \cdot 34.663}{10}\right)^2 \approx 46.158 \Rightarrow 47 \text{ people}$$

7. $n = \left(\frac{z_c \sigma}{E}\right)^2 = \left(\frac{(1.96)(7.098)}{2}\right)^2 \approx 48.39 \Rightarrow 49 \text{ people}$

9. $t_c = 2.365$ **11.** $t_c = 2.624$

13. $E = t_c \dfrac{s}{\sqrt{n}} = 1.753 \dfrac{25.6}{\sqrt{16}} \approx 11.2$

15. $E = t_c \dfrac{s}{\sqrt{n}} = 2.718\left(\dfrac{0.9}{\sqrt{12}}\right) \approx 0.7$

17. $\bar{x} \pm z_c \dfrac{s}{\sqrt{n}} = 72.1 \pm 1.753 \dfrac{25.6}{\sqrt{16}} = 72.1 \pm 11.219 \approx (60.9, 83.3)$

19. $\bar{x} \pm t_c \dfrac{s}{\sqrt{n}} = 6.8 \pm 2.718\left(\dfrac{0.9}{\sqrt{12}}\right) = 6.8 \pm 0.706 = (6.1, 7.5)$

21. $\bar{x} \pm t_c \dfrac{s}{\sqrt{n}} = 80 \pm 1.761 \dfrac{14}{\sqrt{15}} = 80 \pm 6.366 \approx (74, 86)$

23. $\hat{p} = \dfrac{x}{n} = \dfrac{560}{2000} = 0.28, \hat{q} = 0.72$ **25.** $\hat{p} = \dfrac{x}{n} = \dfrac{442}{2010} = 0.220, \hat{q} = 0.780$

27. $\hat{p} = \dfrac{x}{n} = \dfrac{116}{644} = 0.180, \hat{q} = 0.820$ **29.** $\hat{p} = \dfrac{x}{n} = \dfrac{2021}{4813} = 0.420, \hat{q} = 0.580$

31. $\hat{p} \pm z_c \sqrt{\dfrac{\hat{p}\hat{q}}{n}} = 0.28 \pm 1.96 \sqrt{\dfrac{0.28 \cdot 0.72}{2000}} = 0.28 \pm 0.020 \approx (0.260, 0.300)$

33. $\hat{p} \pm z_c \sqrt{\dfrac{\hat{p}\hat{q}}{n}} = 0.220 \pm 1.645 \sqrt{\dfrac{0.220 \cdot 0.780}{2010}} = 0.220 \pm 0.015 \approx (0.205, 0.235)$

35. $\hat{p} \pm z_c \sqrt{\dfrac{\hat{p}\hat{q}}{n}} = 0.180 \pm 2.575 \sqrt{\dfrac{0.180 \cdot 0.820}{644}} = 0.180 \pm 0.039 = (0.141, 0.219)$

37. $\hat{p} \pm z_c \sqrt{\dfrac{\hat{p}\hat{q}}{n}} = 0.420 \pm 1.96 \sqrt{\dfrac{(0.420)(0.580)}{4813}} = 0.420 \pm 0.014 = (0.406, 0.434)$

39. (a) $n = \hat{p}\hat{q}\left(\dfrac{z_c}{E}\right)^2 = 0.50 \cdot 0.50\left(\dfrac{1.96}{0.05}\right)^2 \approx 384.16 \rightarrow 385$ adults

(b) $n = \hat{p}\hat{q}\left(\dfrac{z_c}{E}\right)^2 = 0.63 \cdot 0.37\left(\dfrac{1.96}{0.05}\right)^2 \approx 358.19 \rightarrow 359$ adults

(c) The minimum sample size needed is smaller when a preliminary estimate is available.

41. $\chi_R^2 = 23.337, \chi_L^2 = 4.404$

43. $\chi_R^2 = 14.067, \chi_L^2 = 2.167$

45. $s = 0.0727$

95% CI for σ^2: $\left(\dfrac{(n-1)s^2}{\chi_R^2}, \dfrac{(n-1)s^2}{\chi_L^2}\right) = \left(\dfrac{15 \cdot (0.0727)^2}{27.488}, \dfrac{15 \cdot (0.0727)^2}{6.262}\right) \approx (0.0029, 0.0127)$

95% CI for σ: $\left(\sqrt{0.00288}, \sqrt{0.01266}\right) \approx (0.0537, 0.1125)$

47. $s = 1.125$

90% CI for σ^2: $\left(\dfrac{(n-1)s^2}{\chi_R^2}, \dfrac{(n-1)s^2}{\chi_L^2}\right) = \left(\dfrac{(23)(1.125)^2}{35.172}, \dfrac{(23)(1.125)^2}{13.091}\right) = (0.83, 2.22)$

90% CI for σ: $\left(\sqrt{0.83}, \sqrt{2.22}\right) \approx (0.91, 1.49)$

CHAPTER 6 QUIZ SOLUTIONS

1. (a) $\bar{x} \approx 98.110$

   (b) $s \approx 24.722$

$$E = t_c \frac{s}{\sqrt{n}} = 1.960 \frac{24.722}{\sqrt{30}} \approx 8.847$$

   (c) $\bar{x} \pm t_c \frac{s}{\sqrt{n}} = 98.110 \pm 1.960 \frac{24.722}{\sqrt{30}} = 98.110 \pm 8.847 \approx (89.263, 106.957)$

   You are 95% confident that the population mean repair costs is contained between \$89.26 and \$106.96.

2. $n = \left(\frac{z_c \sigma}{E}\right)^2 = \left(\frac{2.575 \cdot 22.50}{10}\right) \approx 33.568 \rightarrow 34$ dishwashers

3. (a) $\bar{x} = 12.96$

   (b) $s \approx 1.35$

   (c) $\bar{x} \pm t_c \frac{s}{\sqrt{n}} = 12.96 \pm 1.833 \frac{1.35}{\sqrt{10}} = 12.96 \pm 0.783 \approx (12.18, 13.74)$

   (d) $\bar{x} \pm z_c \frac{\sigma}{\sqrt{n}} = 12.96 \pm 1.645 \frac{2.63}{\sqrt{10}} = 12.96 \pm 1.368 \approx (11.59, 14.33)$

   The z-CI is wider because $\sigma = 2.63$ while $s = 1.35$.

4. $\bar{x} \pm t_c \frac{s}{\sqrt{n}} = 6824 \pm 2.447 \frac{340}{\sqrt{7}} = 6824 \pm 314.46 \approx (6510, 7138)$

5. (a) $\hat{p} = \frac{x}{n} = \frac{643}{1037} = 0.620$

   (b) $\hat{p} \pm z_c \sqrt{\frac{\hat{p}\hat{q}}{n}} = 0.620 \pm 1.645 \sqrt{\frac{0.620 \cdot 0.38}{1037}} = 0.620 \pm 0.025 \approx (0.595, 0.645)$

   (c) $n = \hat{p}\hat{q}\left(\frac{z_c}{E}\right)^2 = 0.620 \cdot 0.38 \left(\frac{2.575}{0.04}\right)^2 \approx 976.36 \rightarrow 977$ adults

Note: The answer for Exercise 6 may differ slightly from the text answer due to rounding.

6. (a) $\left(\frac{(n-1)s^2}{\chi_R^2}, \frac{(n-1)s^2}{\chi_L^2}\right) = \left(\frac{29 \cdot (24.722)^2}{45.722}, \frac{29 \cdot (24.722)^2}{16.047}\right) \approx (387.650, 1104.514)$

   (b) $\left(\sqrt{387.650}, \sqrt{1104.514}\right) \approx (19.689, 33.234)$

Hypothesis Testing with One Sample

7.1 Try It Yourself Solutions

1a. (1) The mean ... is not 74 months.

$\mu \neq 74$

(2) The variance ... is less than or equal to 3.5.

$\sigma^2 \leq 3.5$

(3) The proportion ... is greater than 39%.

$p > 0.39$

b. (1) $\mu = 74$ (2) $\sigma^2 > 3.5$ (3) $p \leq 0.39$

c. (1) $H_0: \mu = 74; H_a: \mu \neq 74;$ (claim)

(2) $H_0: \sigma^2 \leq 3.5$ (claim); $H_a: \sigma^2 > 3.5$

(3) $H_0: p \leq 0.39; H_a: p > 0.39$ (claim)

2a. $H_0: p \leq 0.01; H_a: p > 0.01$

b. Type I error will occur if the actual proportion is less than or equal to 0.01, but you reject H_0.

Type II error will occur if the actual proportion is greater than 0.01, but you fail to reject H_0.

c. Type II error is more serious because you would be misleading the consumer, possibly causing serious injury or death.

3a. (1) $H_0: \mu = 74; H_a: \mu \neq 74$

(2) $H_0: p \leq 0.39; H_a: p > 0.39$

b. (1) Two-tailed   (2) Right-tailed

c. (1) 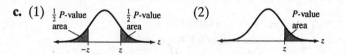 (2)

4a. There is enough evidence to support the radio station's claim.

b. There is not enough evidence to support the radio station's claim.

5a. (1) Support claim.   (2) Reject claim.

b. (1) $H_0: \mu \geq 650; H_a: \mu < 650$ (claim)

(2) $H_0: \mu = 98.6$ (claim); $H_a: \mu \neq 98.6$

7.1 EXERCISE SOLUTIONS

1. Null hypothesis (H_0) and alternative hypothesis (H_a). One represents the claim, the other, its complement.

3. False. In a hypothesis test, you assume the null hypothesis is true.

5. True

7. False. A small P-value in a test will favor a rejection of the null hypothesis.

9. $H_0: \mu \leq 645$ (claim); $H_a: \mu > 645$

11. $H_0: \sigma = 5$; $H_a: \sigma \neq 5$ (claim)

13. $H_0: p \geq 0.45$; $H_a: p < 0.45$ (claim)

15. c, $H_0: \mu \leq 3$

17. b, $H_0: \mu = 3$

19. Right-tailed

21. Two-tailed

23. $\mu > 750$

$H_0: \mu \leq 750$; $H_a: \mu > 750$ (claim)

25. $\sigma \leq 320$

$H_0: \sigma \leq 320$ (claim); $H_a: \sigma > 320$

27. $p = 0.81$

$H_0: p = 0.81$ (claim); $H_a: p \neq 0.81$

29. Type I: Rejecting $H_0: p \geq 0.60$ when actually $p \geq 0.60$.

Type II: Not rejecting $H_0: p \geq 0.60$ when actually $p < 0.60$.

31. Type I: Rejecting $H_0: \sigma \leq 12$ when actually $\sigma \leq 12$.

Type II: Not rejecting $H_0: \sigma \leq 12$ when actually $\sigma > 12$.

33. Type I: Rejecting $H_0: p = 0.88$ when actually $p = 0.88$.

Type II: Not rejecting $H_0: p = 0.88$ when actually $p \neq 0.88$.

35. The null hypothesis is $H_0: p \geq 0.14$, the alternative hypothesis is $H_a: p < 0.14$. Therefore, because the alternative hypothesis contains <, the test is a left-tailed test.

37. The null hypothesis is $H_0: p = 0.87$, the alternative hypothesis is $H_a: p \neq 0.87$. Therefore, because the alternative hypothesis contains ≠, the test is a two-tailed test.

39. The null hypothesis is $H_0: p = 0.053$, the alternative hypothesis is $H_a: p \neq 0.053$. Therefore, because the alternative hypothesis contains ≠, the test is a two-tailed test.

41. (a) There is enough evidence to support the company's claim.

(b) There is not enough evidence to support the company's claim.

43. (a) There is enough evidence to support the Department of Labor's claim.

(b) There is not enough evidence to support the Department of Labor's claim.

45. (a) There is enough evidence to support the manufacturer's claim.

(b) There is not enough evidence to support the manufacturer's claim.

47. $H_0: \mu \geq 60$; $H_a: \mu < 60$

49. (a) $H_0: \mu \geq 15; H_a: \mu < 15$

(b) $H_0: \mu \leq 15\ H_a: \mu > 15$

51. If you decrease α, you are decreasing the probability that you reject H_0. Therefore, you are increasing the probability of failing to reject H_0. This could increase β, the probability of failing to reject H_0 when H_0 is false.

53. (a) Fail to reject H_0 because the CI includes values greater than 70.

(b) Reject H_0 because the CI is located below 70.

(c) Fail to reject H_0 because the CI includes values greater than 70.

55. (a) Reject H_0 because the CI is located to the right of 0.20.

(b) Fail to reject H_0 because the CI includes values less than 0.20.

(c) Fail to reject H_0 because the CI includes values less than 0.20.

7.2 HYPOTHESIS TESTING FOR THE MEAN (LARGE SAMPLES)

7.2 Try It Yourself Solutions

1a. (1) $P = 0.0347 > 0.01 = \alpha$

(2) $P = 0.0347 < 0.05 = \alpha$

b. (1) Fail to reject H_0 because $0.0347 > 0.01$.

(2) Reject H_0 because $0.0347 < 0.05$.

2a.

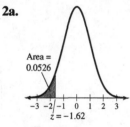

b. $P = 0.0526$

c. Fail to reject H_0 because $P = 0.0526 > 0.05 = \alpha$.

3a. Area that corresponds to $z = 2.31$ is 0.9896.

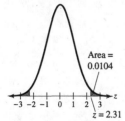

b. $P = 2\ (\text{area}) = 2(0.0104) = 0.0208$

c. Fail to reject H_0 because $P = 0.0208 > 0.01 = \alpha$.

4a. The claim is "the mean speed is greater than 35 miles per hour."

$H_0: \mu \le 35; H_a: \mu > 35$ (claim)

b. $\alpha = 0.05$

c. $z = \dfrac{\bar{x} - \mu}{\dfrac{s}{\sqrt{n}}} = \dfrac{36 - 35}{\dfrac{4}{\sqrt{100}}} = \dfrac{1}{0.4} = 2.500$

d. P-value = Area right of $z = 2.50 = 0.0062$

e. Reject H_0 because P-value $= 0.0062 < 0.05 = \alpha$.

f. Because you reject H_0, there is enough evidence to claim the average speed limit is greater than 35 miles per hour.

5a. The claim is "one of your distributors reports an average of 150 sales per day."

$H_0: \mu = 150$ (claim); $H_a: \mu \ne 150$

b. $\alpha = 0.01$

c. $z = \dfrac{\bar{x} - \mu}{\dfrac{\sigma}{\sqrt{n}}} = \dfrac{143 - 150}{\dfrac{15}{\sqrt{35}}} = \dfrac{-7}{2.535} \approx -2.76$

d. P-value $= 0.0058$

e. Reject H_0 because P-value $= 0.0058 < 0.01 = \alpha$.

f. There is enough evidence to reject the claim.

6a. $P = 0.0440 > 0.01 = \alpha$ **b.** Fail to reject H_0.

7a.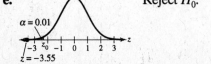

8a.

b. Area $= 0.1003$

b. 0.0401 and 0.9599

c. $z_0 = -1.28$

c. $z_0 = -1.75$ and 1.75

d. $z < -1.28$

d. $z < -1.75, z > 1.75$

9a. The claim is "the mean work day of the firm's accountants is less than 8.5 hours."

$H_0: \mu > 8.5; H_a: \mu < 8.5$ (claim)

b. $\alpha = 0.01$

c. $z_0 = -2.33$; Rejection region: $z < -2.33$

d. $z = \dfrac{x - \mu}{\dfrac{s}{\sqrt{n}}} = \dfrac{8.2 - 8.5}{\dfrac{0.5}{\sqrt{35}}} = \dfrac{-0.300}{0.0845} \approx -3.550$

e. Reject H_0.

$z = -3.55$

f. There is enough evidence to support the claim.

10a. $\alpha = 0.01$

   b. $\pm z_0 = \pm 2.575$; Rejection regions: $z < -2.575$, $z > 2.575$

   c. Fail to reject H_0.

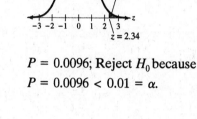

   d. There is not enough evidence to support the claim that the mean cost is significantly different from \$10,460 at the 1% level of significance.

7.2 EXERCISE SOLUTIONS

1.

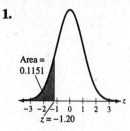

$P = 0.1151$; Fail to reject H_0 because $P = 0.1151 > 0.10 = \alpha$.

3.

$P = 0.0096$; Reject H_0 because $P = 0.0096 < 0.01 = \alpha$.

5.

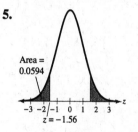

$P = 2(\text{Area}) = 2(0.0594) = 0.1188$;
Fail to reject H_0 because $P = 0.1188 > 0.05 = \alpha$.

7. c   **9.** e   **11.** b

13. (a) Fail to reject H_0.

   (b) Reject H_0 $(P = 0.0461 < 0.05 = \alpha)$.

15. 1.645   **17.** -1.88 **19.** ± 2.33

21. Right-tailed $(\alpha = 0.01)$

23. Two-tailed $(\alpha = 0.10)$

25. (a) Fail to reject H_0 because $-1.645 < z < 1.645$.

   (b) Reject H_0 because $z > 1.645$.

   (c) Fail to reject H_0 because $-1.645 < z < 1.645$.

   (d) Reject H_0 because $z < -1.645$.

27. (a) Fail to reject H_0 because $z < 1.285$.

(b) Fail to reject H_0 because $z < 1.285$.

(c) Fail to reject H_0 because $z < 1.285$.

(d) Reject H_0 because $z > 1.285$.

29. H_0: $\mu = 40$; H_a: $\mu \neq 40$

$\mu = 0.05 \rightarrow z_0 = \pm 1.96$

$$z = \frac{\bar{x} - \mu}{\frac{s}{\sqrt{n}}} = \frac{39.2 - 40}{\frac{3.23}{\sqrt{75}}} = \frac{-0.8}{0.373} \approx -2.145$$

Reject H_0. There is enough evidence to reject the claim.

31. H_0: $\mu = 6000$; H_a: $\mu \neq 6000$

$\alpha = 0.01 \rightarrow z_0 = \pm 2.575$

$$z = \frac{\bar{x} - \mu}{\frac{s}{\sqrt{n}}} = \frac{5800 - 6000}{\frac{350}{\sqrt{35}}} = \frac{-200}{59.161} \approx -3.381$$

Reject H_0. There is enough evidence to support the claim.

33. (a) H_0: $\mu \leq 275$; H_a: $\mu > 275$ (claim)

(b) $z = \dfrac{\bar{x} - \mu}{\frac{s}{\sqrt{n}}} = \dfrac{282 - 275}{\frac{35}{\sqrt{85}}} = \dfrac{7}{3796} \approx 1.84$ Area $= 0.9671$

(c) P-value $= \{$Area to right of $z = 1.84\} = 0.0329$

(d) Reject H_0. There is sufficient evidence at the 4% level of significance to support the claim that the mean score for Illinois' eighth grades is more than 275.

35. (a) H_0: $\mu \leq 8$; H_a: $\mu > 8$ (claim)

(b) $z = \dfrac{\bar{x} - \mu}{\frac{s}{\sqrt{n}}} = \dfrac{7.9 - 8}{\frac{2.67}{\sqrt{100}}} = \dfrac{-0.1}{0.267} \approx -0.37$ Area $= 0.3557$

(c) P-value $= \{$Area to right of $z = -0.37\} = 0.6443$

(d) Fail to reject H_0.

(e) There is insufficient evidence at the 7% level of significance to support the claim that the mean consumption of tea by a person in the United States is more than 8 gallons per year.

37. (a) $H_0: \mu = 15$ (claim); $H_a: \mu \neq 15$

(b) $\bar{x} \approx 14.834$ $s \approx 4.288$

$$z = \frac{\bar{x} - \mu}{\frac{s}{\sqrt{n}}} = \frac{14.834 - 15}{\frac{4.288}{\sqrt{32}}} = \frac{-0.166}{0.758} \approx -0.219 \quad \text{Area} = 0.4129$$

(c) P-value $= 2\{\text{Area to left of } z = -0.22\} = 2\{0.4129\} = 0.8258$

(d) Fail to reject H_0.

(e) There is insufficient evidence at the 5% level of significance to reject the claim that the mean time it takes smokers to quit smoking permanently is 15 years.

39. (a) $H_0: \mu = 40$ (claim); $H_a: \mu \neq 40$

(b) $z_0 = \pm 2.575$;

Rejection regions: $z < -2.575$ and $z > 2.575$

(c) $z = \frac{\bar{x} - \mu}{\frac{s}{\sqrt{n}}} = \frac{39.2 - 40}{\frac{7.5}{\sqrt{30}}} = \frac{-0.8}{1.369} \approx -0.584$

(d) Fail to reject H_0.

(e) There is insufficient evidence at the 1% level of significance to reject the claim that the mean caffeine content per one 12-ounce bottle of cola is 40 milligrams.

41. (a) $H_0: \mu \geq 750$ (claim); $H_a: \mu < 750$

(b) $z_0 = -2.05$; Rejection region: $z < -2.05$

(c) $z = \frac{\bar{x} - \mu}{\frac{s}{\sqrt{n}}} = \frac{745 - 750}{\frac{60}{\sqrt{36}}} = \frac{-5}{10} \approx -0.500$

(d) Fail to reject H_0.

(e) There is insufficient evidence at the 2% level of significance to reject the claim that the mean life of the bulb is at least 750 hours.

43. (a) $H_0: \mu \leq 32$; $H_a: \mu > 32$ (claim)

(b) $z_0 = 1.55$; Rejection region: $z > 1.55$

(c) $\bar{x} \approx 29.676$ $s \approx 9.164$

$$z = \frac{\bar{x} - \mu}{\frac{s}{\sqrt{n}}} = \frac{29.676 - 32}{\frac{9.164}{\sqrt{34}}} = \frac{-2.324}{1.572} \approx -1.478$$

(d) Fail to reject H_0.

(e) There is insufficient evidence at the 6% level of significance to support the claim that the mean nitrogen dioxide level in Calgary is greater than 32 parts per billion.

45. (a) $H_0: \mu \geq 10$ (claim); $H_a: \mu < 10$

(b) $z_0 = -1.88$; Rejection region: $z < -1.88$

(c) $\bar{x} \approx 9.780$, $s \approx 2.362$

$$x = \frac{\bar{x} - \mu}{\frac{s}{\sqrt{n}}} = \frac{9.780 - 10}{\frac{2.362}{\sqrt{30}}} = \frac{-0.22}{0.431} \approx -0.51$$

(d) Fail to reject H_0.

(e) There is insufficient evidence at the 3% level to reject the claim that the mean weight loss after 1 month is at least 10 pounds.

47. $z = \dfrac{\bar{x} - \mu}{\frac{s}{\sqrt{n}}} = \dfrac{11{,}400 - 11{,}500}{\frac{320}{\sqrt{30}}} = \dfrac{-100}{58.424} \approx -1.71$

P-value = {Area left of $z = -1.71$} = 0.0436
Fail to reject H_0 because the standardized test statistic $z = -1.71$ is greater than the critical value $z_0 = -2.33$.

49. (a) $\alpha = 0.02$; Fail to reject H_0.

(b) $\alpha = 0.05$; Reject H_0.

(c) $z = \dfrac{\bar{x} - \mu}{\frac{s}{\sqrt{n}}} = \dfrac{11{,}400 - 11{,}500}{\frac{320}{\sqrt{50}}} = \dfrac{-100}{45.254} \approx -2.21$

P-value = {Area left of $z = -2.21$} = 0.0136 $\rightarrow$ Fail to reject H_0.

(d) $z = \dfrac{\bar{x} - \mu}{\frac{s}{\sqrt{n}}} = \dfrac{11{,}400 - 11{,}500}{\frac{320}{\sqrt{100}}} = \dfrac{-100}{32} \approx -3.13$

P-value = {Area left of $z = -3.13$} < 0.0009 $\rightarrow$ Reject H_0.

51. Using the classical z-test, the test statistic is compared to critical values. The z-test using a P-value compares the P-value to the level of significance α.

7.3 HYPOTHESIS TESTING FOR THE MEAN (SMALL SAMPLES)

7.3 Try It Yourself Solutions

1a. 2.650   **b.** $t_0 = -2.650$

2a. 1.860   **b.** $t_0 = +1.860$

3a. 2.947   **b.** $t_0 = \pm 2.947$

4a. The claim is "the mean cost of insuring a 2005 Honda Pilot LX is at least \$1350."

$H_0: \mu \geq \$1350$ (claim); $H_a: \mu < \$1350$

b. $\alpha = 0.01$ and d.f. $= n - 1 = 8$

c. $t_0 = -2.896$; Reject H_0 if $t \le -2.896$.

d. $t = \dfrac{\bar{x} - \mu}{\dfrac{s}{\sqrt{n}}} = \dfrac{1290 - 1350}{\dfrac{70}{\sqrt{9}}} = \dfrac{-60}{23.333} \approx -2.571$

e. Fail to reject H_0.

f. There is not enough evidence to reject the claim.

5a. The claim is "the mean conductivity of the river is 1890 milligrams per liter."

$H_0: \mu = 1890$ (claim); $H_a: \mu \ne 1890$

b. $\alpha = 0.01$ and d.f. $= n - 1 = 18$

c. $t_0 = \pm 2.878$; Reject H_0 if $t < -2.878$ or $t > 2.878$.

d. $t = \dfrac{\bar{x} - \mu}{\dfrac{s}{\sqrt{n}}} = \dfrac{2500 - 1890}{\dfrac{700}{\sqrt{19}}} = \dfrac{610}{160.591} \approx 3.798$

e. Reject H_0.

f. There is enough evidence to reject the company's claim.

6a. $t = \dfrac{\bar{x} - \mu}{\dfrac{s}{\sqrt{n}}} = \dfrac{172 - 185}{\dfrac{15}{\sqrt{6}}} = \dfrac{-13}{6.124} \approx -2.123$

$P\text{-value} = \{\text{Area left of } t = -2.123\} \approx 0.0436$

b. $P\text{-value} = 0.0436 < 0.05 = \alpha$

c. Reject H_0.

d. There is enough evidence to reject the claim.

7.3 EXERCISE SOLUTIONS

1. Identify the level of significance α and the degrees of freedom, d.f. $= n - 1$. Find the critical value(s) using the t-distribution table in the row with $n - 1$ d.f. If the hypothesis test is:

(1) Left-tailed, use "One Tail, α" column with a negative sign.

(2) Right-tailed, use "One Tail, α" column with a positive sign.

(3) Two-tailed, use "Two Tail, α" column with a negative and a positive sign.

3. $t_0 = 1.717$ **5.** $t_0 = -2.101$

7. $t_0 = \pm 2.779$ **9.** 1.328

11. -2.473 **13.** ± 3.747

15. (a) Fail to reject H_0 because $t > -2.086$.

   (b) Fail to reject H_0 because $t > -2.086$.

   (c) Fail to reject H_0 because $t > -2.086$.

   (d) Reject H_0 because $t < -2.086$.

17. (a) Fail to reject H_0 because $-2.602 < t < 2.602$.

   (b) Fail to reject H_0 because $-2.602 < t < 2.602$.

   (c) Reject H_0 because $t > 2.602$.

   (d) Reject H_0 because $t < -2.602$.

19. $H_0: \mu = 15$ (claim); $H_a: \mu \neq 15$

   $\alpha = 0.01$ and d.f. $= n - 1 = 5$

   $t_0 = \pm 4.032$

   $$t = \frac{\bar{x} - \mu}{\frac{s}{\sqrt{n}}} = \frac{13.9 - 15}{\frac{3.23}{\sqrt{6}}} = \frac{-1.1}{1.319} \approx -0.834$$

   Fail to reject H_0. There is not enough evidence to reject the claim.

21. $H_0: \mu \geq 8000$ (claim); $H_a: \mu < 8000$

   $\alpha = 0.01$ and d.f. $= n - 1 = 24$

   $t_0 = -2.492$

   $$t = \frac{\bar{x} - \mu}{\frac{s}{\sqrt{n}}} = \frac{7700 - 8000}{\frac{450}{\sqrt{25}}} = \frac{-300}{90} \approx -3.333$$

   Reject H_0. There is enough evidence to reject the claim.

23. (a) $H_0: \mu > 100$; $H_a: \mu < 100$ (claim)

   (b) $t_0 = -3.747$; Reject H_0 if $t < -3.747$.

   (c) $t = \dfrac{\bar{x} - \mu}{\frac{s}{\sqrt{n}}} = \dfrac{75 - 100}{\frac{12.50}{\sqrt{5}}} = \dfrac{-25}{5.590} \approx -4.472$

   (d) Reject H_0.

   (e) There is sufficient evidence at the 1% significance level to support the claim that the mean repair cost for damaged microwave ovens is less than $100.

25. (a) $H_0: \mu \le 1$; $H_a: \mu > 1$ (claim)

  (b) $t_0 = 1.796$; Reject H_0 if $t > 1.796$.

  (c) $t = \dfrac{\bar{x} - \mu}{\frac{s}{\sqrt{n}}} = \dfrac{1.46 - 1}{\frac{0.28}{\sqrt{12}}} = \dfrac{0.46}{0.081} \approx 5.691$

  (d) Reject H_0.

  (e) There is sufficient evidence at the 5% significance level to support the claim that the mean waste recycled by adults in the United States is more than 1 pound per person per day.

27. (a) $H_0: \mu = \$25,000$ (claim); $H_a: \mu \ne \$25,000$

  (b) $t_0 = \pm 2.262$; Reject H_0 if $t < -2.262$ or $t > 2.262$.

  (c) $\bar{x} \approx 25,852.2$ $s \approx \$3197.1$

  $t = \dfrac{\bar{x} - \mu}{\frac{s}{\sqrt{n}}} = \dfrac{25,852.2 - 25,000}{\frac{3197.1}{\sqrt{10}}} = \dfrac{-852.2}{1011.0} \approx 0.843$

  (d) Fail to reject H_0.

  (e) There is insufficient evidence at the 5% significance level to reject the claim that the mean salary for full-time male workers over age 25 without a high school diploma is $25,000.

29. (a) $H_0: \mu \ge 3.0$; $H_a: \mu < 3.0$ (claim)

  (b) $\bar{x} = 1.925$ $x = 0.654$

  $t = \dfrac{\bar{x} - \mu}{\frac{s}{\sqrt{n}}} = \dfrac{1.925 - 3.0}{\frac{0.654}{\sqrt{20}}} = \dfrac{-1.075}{0.146} \approx -7.351$

  P-value $= \{$Area left of $t = -7.351\} \approx 0$

  (c) Reject H_0.

  (e) There is sufficient evidence at the 5% significance level to support the claim that teenage males drink fewer than three 12-ounce servings of soda per day.

31. (a) $H_0: \mu \ge 32$; $H_a: \mu < 32$ (claim)

  (b) $\bar{x} = 30.167$ $s = 4.004$

  $t = \dfrac{\bar{x} - \mu}{\frac{s}{\sqrt{n}}} = \dfrac{30.167 - 32}{\frac{4.004}{\sqrt{18}}} = \dfrac{-1.833}{0.944} \approx -1.942$

  P-value $= \{$Area left of $t = -1.942\} \approx 0.0344$

  (c) Fail to reject H_0.

  (e) There is insufficient evidence at the 1% significance level to support the claim that the mean class size for full-time faculty is fewer than 32.

33. (a) H_0: $\mu = \$2634$ (claim); H_a: $\mu \neq \$2634$

(b) $\bar{x} = \$2785.6$ $s = \$759.3$

$$t = \frac{\bar{x} - \mu}{\frac{s}{\sqrt{n}}} = \frac{2785.6 - 2634}{\frac{759.3}{\sqrt{12}}} = \frac{151.6}{219.19} = 0.692$$

P-value = $2\{$Area right of $t = 0.692\} = 2\{0.2518\} = 0.5036$

(c) Fail to reject H_0.

(e) There is insufficient evidence at the 2% significance level to reject the claim that the typical household in the U.S. spends a mean amount of $2634 per year on food away from home.

35. H_0: $\mu \leq \$2328$; H_a: $\mu > 2328$ (claim)

$$t = \frac{\bar{x} - \mu}{\frac{s}{\sqrt{n}}} = \frac{2528 - 2328}{\frac{325}{\sqrt{6}}} = \frac{200}{132.681} \approx 1.507$$

P-value = $\{$Area right of t = 1.507$\} \approx 0.096$

Because $0.096 > 0.01 = \alpha$, fail to reject H_0.

37. Because σ is unknown, $n < 30$, and the gas mileage is normally distributed, use the t-distribution.

H_0: $\mu \geq 23$ (claim); H_a: $\mu < 23$

$$t = \frac{\bar{x} - \mu}{\frac{s}{\sqrt{n}}} = \frac{22 - 23}{\frac{4}{\sqrt{5}}} = \frac{-1}{1.789} \approx -0.559$$

P-value = $\{$Area left of $t = -0.559\} = 0.303$

Fail to reject H_0. There is insufficient evidence at the 5% significance level to reject the claim that the mean gas mileage for the luxury sedan is at least 23 miles per gallon.

7.4 HYPOTHESIS TESTING FOR PROPORTIONS

7.4 Try It Yourself Solutions

1a. $np = (86)(0.30) = 25.8 > 5$, $nq = (86)(0.70) = 60.2 > 5$

b. The claim is "less than 30% of cellular phone users whose phone can connect to the Internet have done so while at home."

H_0: $p \geq 0.30$; H_a: $p < 0.30$ (claim)

c. $\alpha = 0.05$

d. $z_0 = -1.645$; Reject H_0 if $z < -1.645$.

e. $z = \dfrac{\hat{p} - p}{\sqrt{\dfrac{pq}{n}}} = \dfrac{0.20 - 0.30}{\sqrt{\dfrac{(0.30)(0.70)}{86}}} = \dfrac{-0.1}{0.0494} \approx -2.024$

f. Reject H_0.

g. There is enough evidence to support the claim.

2a. $np = (250)(0.05) = 12.5 > 5, nq = (250)(0.95) = 237.5 > 5$

b. The claim is "5% of U.S. adults have had vivid dreams about UFOs."

$H_0: p = 0.05$ (claim); $H_a: p \neq 0.05$

c. $\alpha = 0.01$

d. $z_0 = \pm 2.575$; Reject H_0 if $z < -2.575$ or $z > 2.575$.

e. $z = \dfrac{\hat{p} - p}{\sqrt{\dfrac{pq}{n}}} = \dfrac{0.08 - 0.05}{\sqrt{\dfrac{(0.05)(0.95)}{250}}} = \dfrac{0.03}{0.0138} \approx 2.176$

f. Fail to reject H_0.

g. There is not enough evidence to reject the claim.

3a. $np = (75)(0.30) = 22.5 > 5, nq = (75)(0.70) = 52.5 > 5$

b. The claim is "more than 30% of U.S. adults regularly watch the Weather Channel."

$H_0: p \leq 0.30$; $H_a: p > 0.30$ (claim)

c. $\alpha = 0.01$

d. $z_0 = 2.33$; Reject H_0 if $z > 2.33$.

e. $\hat{p} = \dfrac{x}{n} = \dfrac{27}{75} = 0.360$

$z = \dfrac{\hat{p} - p}{\sqrt{\dfrac{pq}{n}}} = \dfrac{0.360 - 0.30}{\sqrt{\dfrac{(0.30)(0.70)}{75}}} = \dfrac{0.06}{0.053} \approx 1.13$

f. Fail to reject H_0.

g. There is not enough evidence to support the claim.

7.4 EXERCISE SOLUTIONS

1. Verify that $np \geq 5$ and $nq \geq 5$. State H_0 and H_a. Specify the level of significance α. Determine the critical value(s) and rejection region(s). Find the standardized test statistic. Make a decision and interpret in the context of the original claim.

3. $np = (105)(0.25) = 26.25 > 5$

$nq = (105)(0.75) = 78.75 > 5 \rightarrow$ use normal distribution

$H_0: p = 0.25$; $H_a: p \neq 0.25$ (claim)

$z_0 \pm 1.96$

$z = \dfrac{\hat{p} - p}{\sqrt{\dfrac{pq}{n}}} = \dfrac{0.239 - 0.25}{\sqrt{\dfrac{(0.25)(0.75)}{105}}} = \dfrac{-0.011}{0.0423} \approx -0.260$

Fail to reject H_0. There is not enough evidence to support the claim.

5. $np = (20)(0.12) = 2.4 < 5$

$nq = (20)(0.88) = 17.6 \geq 5 \rightarrow$ cannot use normal distribution

7. $np = (70)(0.48) = 33.6 \geq 5$

$nq = (70)(0.52) = 36.4 \geq 5 \rightarrow$ use normal distribution

H_0: $p \geq 0.48$ (claim); H_a: $p < 0.48$

$z_0 = -1.29$

$$z = \frac{\hat{p} - p}{\sqrt{\dfrac{pz}{n}}} = \frac{0.40 - 0.48}{\sqrt{\dfrac{(0.48)(0.52)}{70}}} = \frac{-0.08}{0.060} \approx -1.34$$

Reject H_0. There is enough evidence to reject the claim.

9. (a) H_0: $p \geq 0.20$ (claim); H_a: $p < 0.20$

(b) $z_0 = -2.33$; Reject H_0 if $z < -2.33$.

(c) $z = \dfrac{\hat{p} - p}{\sqrt{\dfrac{pq}{n}}} = \dfrac{0.185 - 0.20}{\sqrt{\dfrac{(0.20)(0.80)}{200}}} = \dfrac{-0.015}{0.0283} \approx -0.53$

(d) Fail to reject H_0.

(e) There is insufficient evidence at the 1% significance level to reject the claim that at least 20% of U.S. adults are smokers.

11. (a) H_0: $p \leq 0.30$; H_a: $p > 0.30$ (claim)

(b) $z_0 = 1.88$; Reject H_0 if $z > 1.88$.

(c) $z = \dfrac{\hat{p} - p}{\sqrt{\dfrac{pq}{n}}} = \dfrac{0.32 - 0.3}{\sqrt{\dfrac{(0.30)(0.70)}{1050}}} = \dfrac{0.02}{0.0141} \approx 1.41$

(d) Fail to reject H_0.

(e) There is insufficient evidence at the 3% significance level to support the claim that more than 30% of U.S. consumers have stopped buying the product because the manufacturing of the product pollutes the environment.

13. (a) H_0: $p = 0.44$ (claim); H_a: $p \neq 0.44$

(b) $z_0 = \pm 2.33$; Reject H_0 if $z < -2.33$ or $z > 2.33$.

(c) $\hat{p} = \dfrac{722}{1762} \approx 0.410$

$$z = \frac{\hat{p} - p}{\sqrt{\dfrac{pq}{n}}} = \frac{0.410 - 0.44}{\sqrt{\dfrac{(0.44)(0.56)}{1762}}} = \frac{-0.03024}{0.01183} \approx -2.537$$

(d) Reject H_0.

(e) There is sufficient evidence at the 2% significance level to reject the claim that 44% of home buyers find their real estate agent through a friend.

15. $H_0: p \geq 0.52$ (claim); $H_a: p < 0.52$

$z_0 = -1.645$; Rejection region: $z < -1.645$

$$z = \frac{\hat{p} - p}{\sqrt{\dfrac{pq}{n}}} = \frac{0.48 - 0.52}{\sqrt{\dfrac{(0.52)(0.48)}{50}}} = \frac{-0.04}{0.0707} \approx -0.566$$

Fail to reject H_0. There is insufficient evidence to reject the claim.

17. $H_0: p = 0.44$ (claim); $H_a: p \neq 0.44$

$$z = \frac{x - np}{\sqrt{npq}} = \frac{722 - (1762)(0.44)}{\sqrt{(1762)(0.44)(0.56)}} = \frac{-53.28}{20.836} \approx -2.56$$

Reject H_0. The results are the same.

7.5 HYPOTHESIS TESTING FOR VARIANCE AND STANDARD DEVIATION

7.5 Try It Yourself Solutions

1a. $\chi_0^2 = 33.409$

2a. $\chi_0^2 = 17.708$

3a. $\chi_R^2 = 31.526$ **b.** $\chi_L^2 = 8.231$

4a. The claim is "the variance of the amount of sports drink in a 12-ounce bottle is no more than 0.40."

$H_0: \sigma^2 \leq 0.40$ (claim); $H_a: \sigma^2 > 0.40$

b. $\alpha = 0.01$ and d.f. $= n - 1 = 30$

c. $\chi_0^2 = 50.892$; Reject H_0 if $\chi^2 > 50.892$.

d. $\chi^2 = \dfrac{(n-1)s^2}{\sigma^2} = \dfrac{(30)(0.75)}{0.40} = 56.250$

e. Reject H_0.

f. There is enough evidence to reject the claim.

5a. The claim is "the standard deviation in the length of response times is less than 3.7 minutes."

$H_0: \sigma \geq 3.7$; $H_a: \sigma < 3.7$ (claim)

b. $\alpha = 0.05$ and d.f. $= n - 1 = 8$

c. $\chi_0^2 = 2.733$; Reject H_0 if $\chi^2 < 2.733$.

d. $\chi^2 = \dfrac{(n-1)s^2}{\sigma^2} = \dfrac{(8)(3.0)^2}{(3.7)^2} \approx 5.259$

e. Fail to reject H_0.

f. There is not enough evidence to support the claim.

6a. The claim is "the variance of the diameters in a certain tire model is 8.6."

$H_0: \sigma^2 = 8.6$ (claim); $H_a: \sigma^2 \neq 8.6$

b. $\alpha = 0.01$ and d.f. $= n - 1 = 9$

c. $\chi_L^2 = 1.735$ and $\chi_R^2 = 23.589$

Reject H_0 if $\chi^2 > 23.589$ or $\chi^2 < 1.735$.

d. $\chi^2 = \dfrac{(n-1)s^2}{\sigma^2} = \dfrac{(9)(4.3)}{(8.6)} = 4.50$

e. Fail to reject H_0.

f. There is not enough evidence to reject the claim.

7.5 EXERCISE SOLUTIONS

1. Specify the level of significance α. Determine the degrees of freedom. Determine the critical values using the χ^2 distribution. If (a) right-tailed test, use the value that corresponds to d.f. and α. (b) left-tailed test, use the value that corresponds to d.f. and $1 - \alpha$; and (c) two-tailed test, use the value that corresponds to d.f. and $\frac{1}{2}\alpha$ and $1 - \frac{1}{2}\alpha$.

3. $\chi_0^2 = 38.885$

5. $\chi_0^2 = 0.872$

7. $\chi_L^2 = 7.261$, $\chi_R^2 = 24.996$

9. (a) Fail to reject H_0.
   (b) Fail to reject H_0.
   (c) Fail to reject H_0.
   (d) Reject H_0.

11. (a) Fail to reject H_0.
    (b) Reject H_0.
    (c) Reject H_0.
    (d) Fail to reject H_0.

13. $H_0: \sigma^2 = 0.52$ (claim); $H_a: \sigma^2 \neq 0.52$

$\chi_L^2 = 7.564$, $\chi_R^2 = 30.191$

$\chi^2 = \dfrac{(n-1)s^2}{\sigma^2} = \dfrac{(17)(0.508)^2}{(0.52)} \approx 16.608$

Fail to reject H_0. There is insufficient evidence to reject the claim.

15. (a) $H_0: \sigma^2 = 3$ (claim); $H_a: \sigma^2 \neq 3$

(b) $\chi_L^2 = 13.844$, $\chi_R^2 = 41.923$; Reject H_0 if $\chi^2 > 41.923$ or $\chi^2 < 13.844$.

(c) $\chi^2 = \dfrac{(n-1)s^2}{\sigma^2} = \dfrac{(26)(2.8)}{3} \approx 24.267$

(d) Fail to reject H_0.

(e) There is insufficient evidence at the 5% level of significance to reject the claim that the variance of the life of the appliances is 3.

17. (a) H_0: $\sigma \geq 36$; H_a: $\sigma < 36$ (claim)

(b) $\chi_0^2 = 13.240$; Reject H_0 if $\chi^2 < 13.240$.

(c) $\chi^2 = \dfrac{(n-1)s^2}{\sigma^2} = \dfrac{(21)(33.4)^2}{(36)^2} \approx 18.076$

(d) Fail to reject H_0.

(e) There is insufficient evidence at the 10% significance level to support the claim that the standard deviation for eighth graders on the examination is less than 36.

19. (a) H_0: $\sigma \leq 0.5$ (claim); H_a: $\sigma > 0.5$

(b) $\chi_0^2 = 33.196$; Reject H_0 if $\chi^2 > 33.196$.

(c) $\chi^2 = \dfrac{(n-1)s^2}{\sigma^2} = \dfrac{(24)(0.7)^2}{(0.5)^2} = 47.04$

(d) Reject H_0.

(e) There is sufficient evidence at the 10% significance level to reject the claim that the standard deviation of waiting times is no more than 0.5 minute.

21. (a) H_0: $\sigma \geq \$3500$; H_a: $\sigma < \$3500$ (claim)

(b) $\chi_0^2 = 18.114$; Reject H_0 if $\chi^2 < 18.114$.

(c) $\chi^2 = \dfrac{(n-1)s^2}{\sigma^2} = \dfrac{(27)(4100)^2}{(3500)^2} \approx 37.051$

(d) Fail to reject H_0.

(e) There is insufficient evidence at the 10% significance level to support the claim that the standard deviation of the total charge for patients involved in a crash where the vehicle struck a construction baracade is less than $3500.

23. (a) H_0: $\sigma \leq \$20,000$; H_a: $\sigma > \$20,000$ (claim)

(b) $\chi_0^2 = 24.996$; Reject H_0 if $\chi^2 > 24.996$.

(c) $s = 20,826.145$

$\chi^2 = \dfrac{(n-1)s^2}{\sigma^2} = \dfrac{(15)(20,826.145)^2}{(20,000)^2} \approx 16.265$

(d) Fail to reject H_0.

(e) There is insufficient evidence at the 5% significance level to support the claim that the standard deviation of the annual salaries for actuaries is more than $20,000.

25. $\chi^2 = 37.051$

P-value = {Area left of $\chi^2 = 37.051$} = 0.9059

Fail to reject H_0 because P-value = 0.9059 > 0.10 = α.

27. $\chi^2 = 16.265$

P-value = {Area right of $\chi^2 = 16.265$} = 0.3647

Fail to reject H_0 because P-value = 0.3647 > 0.05 = α.

CHAPTER 7 REVIEW EXERCISE SOLUTIONS

1. $H_0: \mu \le 1479$ (claim); $H_a: \mu > 1479$

3. $H_0: p \ge 0.205$; $H_a: p < 0.205$ (claim)

5. $H_0: \sigma \le 6.2$; $H_a: \sigma > 6.2$ (claim)

7. (a) $H_0: p = 0.73$ (claim); $H_a: p \ne 0.73$

   (b) Type I error will occur if H_0 is rejected when the actual proportion of college students that occasionally or frequently come late to class is 0.63.

   Type II error if H_0 is not rejected when the actual proportion of college students that occasionally or frequently come late to class is not 0.63.

   (c) Two-tailed, because hypothesis compares "= vs ≠".

   (d) There is enough evidence to reject the claim.

   (e) There is not enough evidence to reject the claim.

9. (a) $H_0: \mu \le 50$ (claim); $H_a: \mu > 50$

   (b) Type I error will occur if H_0 is rejected when the actual standard deviation sodium content is no more than 50 milligrams.

   Type II error if H_0 is not rejected when the actual standard deviation sodium content is more than 50 milligrams.

   (c) Right-tailed, because hypothesis compares "≤ vs >".

   (d) There is enough evidence to reject the claim.

   (e) There is not enough evidence to reject the claim.

11. $z_0 \approx -2.05$

13. $z_0 = 1.96$

15. $H_0: \mu \le 45$ (claim); $H_a: \mu > 45$

    $z_0 = 1.645$

    $$z = \frac{\bar{x} - \mu}{\frac{s}{\sqrt{n}}} = \frac{47.2 - 45}{\frac{6.7}{\sqrt{42}}} = \frac{2.2}{1.0338} \approx 2.128$$

    Reject H_0. There is enough evidence to reject the claim.

17. $H_0: \mu \ge 5.500$; $H_a: \mu < 5.500$ (claim)

    $z_0 = -2.33$

    $$z = \frac{\bar{x} - \mu}{\frac{s}{\sqrt{n}}} = \frac{5.497 - 5.500}{\frac{0.011}{\sqrt{36}}} = \frac{-0.003}{0.00183} \approx -1.636$$

    Fail to reject H_0. There is not enough evidence to support the claim.

19. $H_0: \mu \leq 0.05$ (claim); $H_a: \mu > 0.05$

$$z = \frac{\bar{x} - \mu}{\frac{s}{\sqrt{n}}} = \frac{0.057 - 0.05}{\frac{0.018}{\sqrt{32}}} = \frac{0.007}{0.00318} \approx 2.20$$

P-value = {Area right of $z = 2.20$} = 0.0139

$\alpha = 0.10$; Reject H_0.

$\alpha = 0.05$; Reject H_0.

$\alpha = 0.01$; Fail to reject H_0.

21. $H_0: \mu = 326$ (claim); $H_a: \mu \neq 326$

$$z = \frac{\bar{x} - \mu}{\frac{s}{\sqrt{n}}} = \frac{318 - 326}{\frac{25}{\sqrt{50}}} = \frac{-8}{3.536} \approx -2.263$$

P-value = 2{Area left of $z = -2.263$} = 2{0.012} = 0.024

Reject H_0. There is sufficient evidence to reject the claim.

23. $t_0 = \pm 2.093$ **25.** $t_0 = -1.345$

27. $H_0: \mu = 95$; $H_a: \mu \neq 95$ (claim)

$t_0 = \pm 2.201$

$$t = \frac{\bar{x} - \mu}{\frac{s}{\sqrt{n}}} = \frac{94.1 - 95}{\frac{1.53}{\sqrt{12}}} = \frac{-0.9}{0.442} \approx -2.038$$

Fail to reject H_0. There is not enough evidence to support the claim.

29. $H_0: \mu \geq 0$ (claim); $H_a: \mu < 0$

$t_0 = -1.341$

$$t = \frac{\bar{x} - \mu}{\frac{s}{\sqrt{n}}} = \frac{-0.45 - 0}{\frac{1.38}{\sqrt{16}}} = \frac{-0.45}{0.345} \approx -1.304$$

Fail to reject H_0. There is not enough evidence to reject the claim.

31. $H_0: \mu \leq 48$ (claim); $H_a: \mu > 48$

$t_0 = 3.148$

$$t = \frac{\bar{x} - \mu}{\frac{s}{\sqrt{n}}} = \frac{52 - 48}{\frac{2.5}{\sqrt{7}}} = \frac{4}{0.945} \approx 4.233$$

Reject H_0. There is enough evidence to reject the claim.

33. $H_0: \mu = \$25$ (claim); $H_a: \mu \neq \$25$

$t_0 = \pm 1.740$

$$t = \frac{\bar{x} - \mu}{\frac{s}{\sqrt{n}}} = \frac{26.25 - 25}{\frac{3.23}{\sqrt{18}}} = \frac{1.25}{0.761} \approx 1.642$$

Fail to reject H_0. There is not enough evidence to reject the claim.

35. $H_0: \mu \geq \$10,200$ (claim); $H_a: \mu < \$10,200$

$t_0 = -2.602$

$\bar{x} = 9895.8 \quad s = 490.88$

$$t = \frac{\bar{x} - \mu}{\frac{s}{\sqrt{n}}} = \frac{9895.8 - 10,200}{\frac{490.88}{\sqrt{16}}} = \frac{-304.2}{122.72} \approx -2.479$$

P-value ≈ 0.0128

Fail to reject H_0. There is not enough evidence to reject the claim.

37. $H_0: p = 0.15$ (claim); $H_a: p \neq 0.15$

$z_0 = \pm 1.96$

$$z = \frac{\hat{p} - p}{\sqrt{\frac{pq}{n}}} = \frac{0.09 - 0.15}{\sqrt{\frac{(0.15)(0.85)}{40}}} = \frac{-0.06}{0.0565} \approx -1.063$$

Fail to reject H_0. There is not enough evidence to reject the claim.

39. Because $np = 3.6$ is less than 5, the normal distribution cannot be used to approximate the binomial distribution.

41. Because $np = 1.2 < 5$, the normal distribution cannot be used to approximate the binomial distribution.

43. $H_0: p = 0.20$; $H_a: p \neq 0.20$ (claim)

$z_0 = \pm 2.575$

$$z = \frac{\hat{p} - p}{\sqrt{\frac{pq}{n}}} = \frac{0.23 - 0.20}{\sqrt{\frac{(0.20)(0.80)}{56}}} = \frac{0.03}{0.0534} \approx 0.561$$

Fail to reject H_0. There is not enough evidence to support the claim.

45. $H_0: p \leq 0.40$; $H_a: p > 0.40$ (claim)

$z_0 = 1.28$

$$\hat{p} = \frac{x}{n} = \frac{1130}{2730} \approx 0.414$$

$$z = \frac{\hat{p} - p}{\sqrt{\frac{pq}{n}}} = \frac{0.414 - 0.40}{\sqrt{\frac{(0.40)(0.60)}{2730}}} = \frac{0.14}{0.0094} \approx 1.493$$

Reject H_0. There is enough evidence to support the claim.

47. $\chi_R^2 = 30.144$

49. $\chi_R^2 = 33.196$

51. $H_0: \sigma^2 \leq 2$; $H_a: \sigma^2 > 2$ (claim)

$\chi_0^2 = 24.769$

$$\chi^2 = \frac{(n-1)s^2}{\sigma^2} = \frac{(17)(2.95)}{(2)} = 25.075$$

Reject H_0. There is enough evidence to support the claim.

53. $H_0: \sigma^2 = 1.25$ (claim); $H_a: \sigma^2 \neq 1.25$

$\chi_L^2 = 0.831$, $\chi_R^2 = 12.833$

$\chi^2 = \dfrac{(n-1)s^2}{\sigma^2} = \dfrac{(5)(1.03)^2}{(1.25)^2} \approx 3.395$

Fail to reject H_0. There is not enough evidence to reject the claim.

55. $H_0: \sigma^2 \leq 0.01$ (claim); $H_a: \sigma^2 > 0.01$

$\chi_0^2 = 49.645$

$\chi^2 = \dfrac{(n-1)s^2}{\sigma^2} = \dfrac{(27)(0.064)}{(0.01)} = 172.800$

Reject H_0. There is enough evidence to reject the claim.

CHAPTER 7 QUIZ SOLUTIONS

1. (a) $H_0: \mu \geq 22$ (claim); $H_a: \mu < 22$

(b) "$\geq$ vs $<$" $\rightarrow$ Left-tailed

σ is unknown and $n \geq 30 \rightarrow z$-test.

(c) $z_0 = -2.05$; Reject H_0 if $z < -2.05$.

(d) $z = \dfrac{\bar{x} - \mu}{\frac{s}{\sqrt{n}}} = \dfrac{21.6 - 22}{\frac{7}{\sqrt{103}}} = \dfrac{-0.4}{0.690} \approx -0.580$

(e) Fail to reject H_0. There is insufficient evidence at the 2% significance level to reject the claim that the mean utilization of fresh citrus fruits by people in the U.S. is at least 22 pounds per year.

2. (a) $H_0: \mu \geq 20$ (claim); $H_a: \mu < 20$

(b) "$\geq$ vs $<$" $\rightarrow$ Left-tailed

σ is unknown, the population is normal, and $n < 30 \rightarrow t$-test.

(c) $t_0 = -1.895$; Reject H_0 if $t < -1.895$.

(d) $z = \dfrac{\bar{x} - \mu}{\frac{s}{\sqrt{n}}} = \dfrac{18 - 20}{\frac{5}{\sqrt{8}}} = \dfrac{-2}{1.768} \approx -1.131$

(e) Fail to reject H_0. There is insufficient evidence at the 5% significance level to reject the claim that the mean gas mileage is at least 20 miles per gallon.

3. (a) $H_0: p \leq 0.10$ (claim); $H_a: p > 0.10$

(b) "$\leq$ vs $>$" $\rightarrow$ Right-tailed

$np \geq 5$ and $nq \geq 5 \rightarrow z$-test

(c) $z_0 = 1.75$; Reject H_0 if $z > 1.75$.

(d) $z = \dfrac{\hat{p} - p}{\sqrt{\dfrac{pq}{n}}} = \dfrac{0.13 - 0.10}{\sqrt{\dfrac{(0.10)(0.90)}{57}}} = \dfrac{0.03}{0.0397} \approx 0.75$

(e) Fail to reject H_0. There is insufficient evidence at the 4% significance level to reject the claim that no more than 10% of microwaves need repair during the first five years of use.

4. (a) $H_0: \sigma = 113$ (claim); $H_a: \sigma \neq 113$

(b) "= vs ≠ " → Two-tailed

Assuming the scores are normally distributed and you are testing the hypothesized standard deviation → χ^2 test.

(c) $\chi_L^2 = 3.565$, $\chi_R^2 = 29.819$; Reject H_0 if $\chi^2 < 3.565$ or if $\chi^2 > 29.819$.

(d) $\chi^2 = \dfrac{(n-1)s^2}{\sigma^2} = \dfrac{(13)(108)^2}{(113)^2} \approx 11.875$

(e) Fail to reject H_0. There is insufficient evidence at the 1% significance level to reject the claim that the standard deviation of the SAT critical reading scores for the state is 105.

5. (a) $H_0: \mu = \$48{,}718$ (claim); $H_a: \mu \neq \$48{,}718$

(b) "= vs ≠" → Two-tailed

σ is unknown, $n < 30$, and assuming the salaries are normally distributed → t-test.

(c) not applicable

(d) $t = \dfrac{\bar{x} - \mu}{\dfrac{s}{\sqrt{n}}} = \dfrac{47{,}164 - 48{,}718}{\dfrac{6500}{\sqrt{12}}} = \dfrac{-1554}{1876.388} \approx -0.828$

P-value $= 2\{$Area left of $t = -0.828\} = 2(0.2126) = 0.4252$

(e) Fail to reject H_0. There is insufficient evidence at the 5% significance level to reject the claim that the mean annual salary for full-time male workers ages 25 to 34 with a bachelor's degree is $48,718.

6. (a) $H_0: \mu = \$201$ (claim); $H_a: \mu \neq \$201$

(b) "= vs ≠" → Two-tailed

σ is unknown, $n \geq 30$ → z-test.

(c) not applicable

(d) $z = \dfrac{\bar{x} - \mu}{\dfrac{s}{\sqrt{n}}} = \dfrac{216 - 201}{\dfrac{30}{\sqrt{35}}} = \dfrac{15}{5.071} \approx 2.958$

P-value $= 2\{$Area right of $z = 2.958\} = 2\{0.0015\} = 0.0030$

(e) Reject H_0. There is sufficient evidence at the 5% significance level to reject the claim that the mean daily cost of meals and lodging for a family of four traveling in Kansas is $201.

8.1 Try It Yourself Solutions

Note: Answers may differ due to rounding.

1. (1) Independent

    (2) Dependent

2a. $H_0: \mu_1 = \mu_2; H_a: \mu_1 \neq \mu_2$ (claim)

  b. $\alpha = 0.01$

  c. $z_0 = \pm 2.575$; Reject H_0 if $z > 2.575$ or $z < -2.575$.

  d. $z = \dfrac{(\bar{x}_1 - \bar{x}_2) - (\mu_1 - \mu_2)}{\sqrt{\dfrac{s_1^2}{n_1} + \dfrac{s_2^2}{n_2}}} = \dfrac{(3900 - 3500) - (0)}{\sqrt{\dfrac{(900)^2}{50} + \dfrac{(500)^2}{50}}} = \dfrac{400}{\sqrt{21200}} \approx 2.747$

  e. Reject H_0.

  f. There is enough evidence to support the claim.

3a. $z = \dfrac{(\bar{x}_1 - \bar{x}_2) - (\mu_1 - \mu_2)}{\sqrt{\dfrac{s_1^2}{n_1} + \dfrac{s_2^2}{n_2}}} = \dfrac{(293 - 286) - (0)}{\sqrt{\dfrac{(24)^2}{150} + \dfrac{(18)^2}{200}}} = \dfrac{7}{\sqrt{5.46}} \approx 3.00$

    $\rightarrow$ P-value = {area right of $z = 3.00$} = 0.0014

  b. Reject H_0. There is enough evidence to support the claim.

8.1 EXERCISE SOLUTIONS

1. Two samples are dependent if each member of one sample corresponds to a member of the other sample. Example: The weights of 22 people before starting an exercise program and the weights of the same 22 people 6 weeks after starting the exercise program.

Two samples are independent if the sample selected from one population is not related to the sample from the second population. Example: The weights of 25 cats and the weights of 25 dogs.

3. Use P-values.

5. Independent because different students were sampled.

7. Dependent because the same adults were sampled.

9. Independent because different boats were sampled.

11. Dependent because the same tire sets were sampled.

13. $H_0: \mu_1 = \mu_2$ (claim); $H_a: \mu_1 \neq \mu_2$

Rejection regions: $z_0 < -1.96$ and $z_0 > 1.96$ (Two-tailed test)

(a) $\bar{x}_1 - \bar{x}_2 = 16 - 14 = 2$

(b) $z = \dfrac{(\bar{x}_1 - \bar{x}_2) - (\mu_1 - \mu_2)}{\sqrt{\dfrac{s_1^2}{n_1} + \dfrac{s_2^2}{n_2}}} = \dfrac{(16 - 14) - (0)}{\sqrt{\dfrac{(1.1)^2}{50} + \dfrac{(1.5)^2}{50}}} = \dfrac{2}{\sqrt{0.0692}} \approx 7.60$

(c) z is in the rejection region because $7.60 > 1.96$.

(d) Reject H_0. There is enough evidence to reject the claim.

15. $H_0: \mu_1 \geq \mu_2$; $H_a: \mu_1 < \mu_2$ (claim)

Rejection region: $z_0 < -2.33$ (Left-tailed test)

(a) $\bar{x}_1 - \bar{x}_2 = 1225 - 1195 = 30$

(b) $z = \dfrac{(\bar{x}_1 - \bar{x}_2) - (\mu_1 - \mu_2)}{\sqrt{\dfrac{s_1^2}{n_1} + \dfrac{s_2^2}{n_2}}} = \dfrac{(1225 - 1195) - (0)}{\sqrt{\dfrac{(75)^2}{35} + \dfrac{(105)^2}{105}}} = \dfrac{30}{\sqrt{265.714}} \approx 1.84$

(c) z is not in the rejection region because $1.84 > -2.330$.

(d) Fail to reject H_0. There is not enough evidence to support the claim.

17. $H_0: \mu_1 \leq \mu_2$; $H_a: \mu_1 > \mu_2$ (claim)

$z_0 = 2.33$; Reject H_0 if $z > 2.33$.

$z = \dfrac{(\bar{x}_1 - \bar{x}_2) - (\mu_1 - \mu_2)}{\sqrt{\dfrac{s_1^2}{n_1} + \dfrac{s_2^2}{n_2}}} = \dfrac{(5.2 - 5.5) - (0)}{\sqrt{\dfrac{(0.2)^2}{45} + \dfrac{(0.3)^2}{37}}} = \dfrac{-.30}{\sqrt{0.00332}} \approx -5.207$

Fail to reject H_0. There is not enough evidence to support the claim.

19. (a) $H_0: \mu_1 = \mu_2$; $H_a: \mu_1 \neq \mu_2$ (claim)

(b) $z_0 = \pm 1.645$; Reject H_0 if $z < -1.645$ or $z > 1.645$.

(c) $z = \dfrac{(\bar{x}_1 - \bar{x}_2) - (\mu_1 - \mu_2)}{\sqrt{\dfrac{s_1^2}{n_1} + \dfrac{s_2^2}{n_2}}} = \dfrac{(42 - 45) - (0)}{\sqrt{\dfrac{(4.7)^2}{35} + \dfrac{(4.3)^2}{35}}} = \dfrac{-3}{\sqrt{1.159}} \approx -2.786$

(d) Reject H_0.

(e) There is sufficient evidence at the 10% significance level to support the claim that the mean braking distance is different for both types of tires.

21. (a) $H_0: \mu_1 \geq \mu_2$; $H_a: \mu_1 < \mu_2$ (claim)

(b) $z_0 = -2.33$; Reject H_0 if $z < -2.33$.

(c) $z = \dfrac{(\bar{x}_1 - \bar{x}_2) - (\mu_1 - \mu_2)}{\sqrt{\dfrac{s_1^2}{n_1} + \dfrac{s_2^2}{n_2}}} = \dfrac{(75 - 80) - (0)}{\sqrt{\dfrac{(12.50)^2}{47} + \dfrac{(20)^2}{55}}} = \dfrac{-5}{\sqrt{10.597}} \approx -1.54$

(d) Fail to reject H_0.

(e) There is insufficient evidence at the 1% significance level to conclude that the repair costs for Model A are lower than for Model B.

23. (a) $H_0: \mu_1 = \mu_2$ (claim); $H_a: \mu_1 \neq \mu_2$

(b) $z_0 = \pm 2.575$; Reject H_0 if $z < -2.575$ or $z > 2.575$.

(c) $z = \dfrac{(\bar{x}_1 - \bar{x}_2) - (\mu_1 - \mu_2)}{\sqrt{\dfrac{s_1^2}{n_1} + \dfrac{s_2^2}{n_2}}} = \dfrac{(21.0 - 20.8) - (0)}{\sqrt{\dfrac{(5.0)^2}{43} + \dfrac{(4.7)^2}{56}}} = \dfrac{0.2}{\sqrt{0.976}} \approx 0.202$

(d) Fail to reject H_0.

(e) There is insufficient evidence at the 1% significance level to reject the claim that the male and female high school students have equal ACT scores.

25. (a) $H_0: \mu_1 = \mu_2$ (claim); $H_a: \mu_1 \neq \mu_2$

(b) $z_0 = \pm 1.645$; Reject H_0 if $z < -1.645$ or $z > 1.645$.

(c) $z = \dfrac{(\bar{x}_1 - \bar{x}_2) - (\mu_1 - \mu_2)}{\sqrt{\dfrac{s_1^2}{n_1} + \dfrac{s_2^2}{n_2}}} = \dfrac{(131 - 136) - (0)}{\sqrt{\dfrac{(26)^2}{35} + \dfrac{(19)^2}{35}}} = \dfrac{-5}{\sqrt{29.629}} \approx -0.919$

(d) Fail to reject H_0.

(e) There is insufficient evidence at the 10% significance level to reject the claim that the lodging cost for a family traveling in North Carolina is the same as South Carolina.

27. (a) $H_0: \mu_1 = \mu_2$ (claim); $H_a: \mu_1 \neq \mu_2$

(b) $z_0 = \pm 1.645$; Reject H_0 if $z < -1.645$ or $z > 1.645$.

(c) $z = \dfrac{(\bar{x}_1 - \bar{x}_2) - (\mu_1 - \mu_2)}{\sqrt{\dfrac{s_1^2}{n_1} + \dfrac{s_2^2}{n_2}}} = \dfrac{(145 - 138) - (0)}{\sqrt{\dfrac{(28)^2}{50} + \dfrac{(24)^2}{50}}} = \dfrac{7}{\sqrt{27.2}} \approx 1.342$

(d) Fail to reject H_0.

(e) There is insufficient evidence at the 10% significance level to reject the claim that the lodging cost for a family traveling in North Carolina is the same as South Carolina. The new samples do not lead to a different conclusion.

29. (a) $H_0: \mu_1 \leq \mu_2$; $H_a: \mu_1 < \mu_2$ (claim)

(b) $z_0 = 1.96$; Reject H_0 if $z > 1.96$.

(c) $\bar{x}_1 \approx 2.130$, $s_1 \approx 0.490$, $n_1 = 30$
$\bar{x}_2 \approx 1.757$, $s_2 \approx 0.470$, $n_2 = 30$

$z = \dfrac{(\bar{x}_1 - \bar{x}_2) - (\mu_1 - \mu_2)}{\sqrt{\dfrac{s_1^2}{n_1} + \dfrac{s_2^2}{n_2}}} = \dfrac{(2.130 - 1.757) - (0)}{\sqrt{\dfrac{(0.490)^2}{30} + \dfrac{(0.470)^2}{30}}} = \dfrac{0.373}{\sqrt{0.0154}} \approx 3.01$

(d) Reject H_0.

(e) At the 2.5% level of significance, there is sufficient evidence to support the claim.

31. (a) $H_0: \mu_1 = \mu_2$ (claim); $H_a: \mu_1 \neq \mu_2$

(b) $z_0 = \pm 2.575$; Reject H_0 if $z < -2.575$ or $z > 2.575$.

(c) $\bar{x}_1 \approx 0.875, s_1 \approx 0.011, n_1 = 35$

$\bar{x}_2 \approx 0.701, s_2 \approx 0.011, n_2 = 35$

$$z = \frac{(\bar{x}_1 - \bar{x}_2) - (\mu_1 - \mu_2)}{\sqrt{\dfrac{s_1^2}{n_1} + \dfrac{s_2^2}{n_2}}} = \frac{(0.875 - 0.701) - (0)}{\sqrt{\dfrac{(0.011)^2}{35} + \dfrac{(0.011)^2}{35}}} = \frac{0.174}{\sqrt{0.0000006914}} \approx 66.172$$

(d) Reject H_0.

(e) At the 1% level of significance, there is sufficient evidence to reject the claim.

33. They are equivalent through algebraic manipulation of the equation.

$\mu_1 = \mu_2 \rightarrow \mu_1 - \mu_2 = 0$

35. $H_0: \mu_1 - \mu_2 = -9$ (claim); $H_a: \mu_1 - \mu_2 \neq -9$

$z_0 = \pm 2.575$; Reject H_0 if $z < -2.575$ or $z > 2.575$.

$$z = \frac{(\bar{x}_1 - \bar{x}_2) - (\mu_1 - \mu_2)}{\sqrt{\dfrac{s_1^2}{n_1} + \dfrac{s_2^2}{n_2}}} = \frac{(11.5 - 20) - (-9)}{\sqrt{\dfrac{(3.8)^2}{70} + \dfrac{(6.7)^2}{65}}} = \frac{0.5}{\sqrt{0.897}} \approx 0.528$$

Fail to reject H_0. There is not enough evidence to reject the claim.

37. $H_0: \mu_1 - \mu_2 \leq 6000$; $H_a: \mu_1 - \mu_2 > 6000$ (claim)

$z_0 = 1.28$; Reject H_0 if $z > 1.28$.

$$z = \frac{(\bar{x}_1 - \bar{x}_2) - (\mu_1 - \mu_2)}{\sqrt{\dfrac{s_1^2}{n_1} + \dfrac{s_2^2}{n_2}}} = \frac{(67,900 - 64,000) - (6000)}{\sqrt{\dfrac{(8875)^2}{45} + \dfrac{(9175)^2}{42}}} = \frac{-2100}{\sqrt{3,754,647.817}} \approx -1.084$$

Fail to reject H_0. There is not enough evidence to support the claim.

39. $(\bar{x}_1 - \bar{x}_2) - z_c \sqrt{\dfrac{s_1^2}{n_1} + \dfrac{s_2^2}{n_2}} < \mu_1 - \mu_2 < (\bar{x}_1 - \bar{x}_2) + z_c \sqrt{\dfrac{s_1^2}{n_1} + \dfrac{s_2^2}{n_2}}$

$(123.1 - 125) - 1.96 \sqrt{\dfrac{(9.9)^2}{269} + \dfrac{(10.1)^2}{268}} < \mu_1 - \mu_2 < (123.1 - 125) + 1.96 \sqrt{\dfrac{(9.9)^2}{269} + \dfrac{(10.1)^2}{268}}$

$-1.9 - 1.96\sqrt{0.745} < \mu_1 - \mu_2 < -1.9 + 1.96\sqrt{0.745}$

$-3.6 < \mu_1 - \mu_2 < -0.2$

41. $H_0: \mu_1 - \mu_2 \geq 0$; $H_a: \mu_1 - \mu_2 < 0$ (claim)

$z_0 = -1.645$; Reject H_0 if $z < -1.645$.

$$z = \frac{(\bar{x}_1 - \bar{x}_2) - (\mu_1 - \mu_2)}{\sqrt{\dfrac{s_1^2}{n_1} + \dfrac{s_2^2}{n_2}}} = \frac{(123.1 - 125) - (0)}{\sqrt{\dfrac{(9.9)^2}{269} + \dfrac{(10.1)^2}{268}}} = \frac{-1.9}{\sqrt{0.745}} \approx -2.20$$

Reject H_0. There is enough evidence to support the claim. I would recommend using the DASH diet and exercise program over the traditional diet and exercise program.

43. $H_0: \mu_1 - \mu_2 \geq 0; H_a: \mu_1 - \mu_2 < 0$ (claim)

The 95% CI for $\mu_1 - \mu_2$ in Exercise 39 contained values greater than or equal to zero and, as found in Exercise 41, there was enough evidence at the 5% level of significance to support the claim. If zero is not contained in the CI for $\mu_1 - \mu_2$, you reject H_0 because the null hypothesis states that $\mu_1 - \mu_2$ is greater than or equal to zero.

8.2 TESTING THE DIFFERENCE BETWEEN MEANS (SMALL INDEPENDENT SAMPLES)

8.2 Try It Yourself Solutions

1a. $H_0: \mu_1 = \mu_2; H_a: \mu_1 \neq \mu_2$ (claim)

b. $\alpha = 0.05$

c. d.f. $= \min\{n_1 - 1, n_2 - 1\} = \min\{8 - 1, 10 - 1\} = 7$

d. $t_0 = \pm 2.365$; Reject H_0 if $t < -2.365$ or $t > 2.365$.

e. $t = \dfrac{(\bar{x}_1 - \bar{x}_2) - (\mu_1 - \mu_2)}{\sqrt{\dfrac{s_1^2}{n_1} + \dfrac{s_2^2}{n_2}}} = \dfrac{(141 - 151) - (0)}{\sqrt{\dfrac{(7.0)^2}{8} + \dfrac{(3.1)^2}{10}}} = \dfrac{-10}{\sqrt{7.086}} \approx -3.757$

f. Reject H_0.

g. There is enough evidence to support the claim.

2a. $H_0: \mu_1 \geq \mu_2; H_a: \mu_1 < \mu_2$ (claim)   **b.** $\alpha = 0.10$

c. d.f. $= n_1 = n_2 - 2 = 12 + 15 - 2 = 25$ **d.** $t_0 = -1.316$; Reject H_0 if $t < -1.316$.

e. $t = \dfrac{(\bar{x}_1 - \bar{x}_2) - (\mu_1 - \mu_2)}{\sqrt{\dfrac{(n_1 - 1)s_1^2 + (n_2 - 1)s_2^2}{n_1 + n_2 - 2}}\sqrt{\dfrac{1}{n_1} + \dfrac{1}{n_2}}} = \dfrac{(32 - 35) - (0)}{\sqrt{\dfrac{(12 - 1)(2.1)^2 + (15 - 1)(1.8)^2}{12 + 15 - 2}}\sqrt{\dfrac{1}{12} + \dfrac{1}{15}}}$

$= \dfrac{-3}{\sqrt{3.755}\,\sqrt{0.15}} \approx -3.997$

f. Reject H_0.

g. There is enough evidence to support the claim.

8.2 EXERCISE SOLUTIONS

1. State hypotheses and identify the claim. Specify the level of significance. Determine the degrees of freedom. Find the critical value(s) and identify the rejection region(s). Find the standardized test statistic. Make a decision and interpret in the context of the original claim.

3. (a) d.f. $= n_1 + n_2 - 2 = 23$
$t_0 = \pm 1.714$

(b) d.f. $= \min\{n_1 - 1, n_2 - 1\} = 10$
$t_0 = \pm 1.812$

5. (a) d.f. $= n_1 + n_2 - 2 = 22$
$t_0 = -2.074$

(b) d.f. $= \min\{n_1 - 1, n_2 - 1\} = 8$
$t_0 = -2.306$

7. (a) d.f. $= n_1 + n_2 - 2 = 19$

$t_0 = 1.729$

(b) d.f. $= \min\{n_1 - 1, n_2 - 1\} = 7$

$t_0 = 1.895$

9. (a) d.f. $= n_1 + n_2 - 2 = 27$

$t_0 = \pm 2.771$

(b) d.f. $= \min\{n_1 - 1, n_2 - 1\} = 11$

$t_0 = \pm 3.106$

11. $H_0\colon \mu_1 = \mu_2$ (claim); $H_a\colon \mu_1 \neq \mu_2$

d.f. $= n_1 + n_2 - 2 = 15$

$t_0 = \pm 2.947$ (Two-tailed test)

(a) $\bar{x}_1 - \bar{x}_2 = 33.7 - 35.5 = -1.8$

(b) $t = \dfrac{(\bar{x}_1 - \bar{x}_2) - (\mu_1 - \mu_2)}{\sqrt{\dfrac{(n_1 - 1)s_1^2 + (n_2 - 1)s_2^2}{n_1 + n_2 - 2}}\sqrt{\dfrac{1}{n_1} + \dfrac{1}{n_2}}} = \dfrac{(33.7 - 35.5) - (0)}{\sqrt{\dfrac{(10 - 1)(3.5)^2 + (7 - 1)(2.2)^2}{10 + 7 - 2}}\sqrt{\dfrac{1}{10} + \dfrac{1}{7}}}$

$= \dfrac{-1.8}{\sqrt{9.286}\,\sqrt{0.243}} \approx -1.199$

(c) t is not in the rejection region.

(d) Fail to reject H_0.

13. There is no need to run the test since this is a right-tailed test and the test statistic is negative. It is obvious that the standardized test statistic will also be negative and fall outside of the rejection region. So, the decision is to fail to reject H_0.

15. (a) $H_0\colon \mu_1 = \mu_2$ (claim); $H_a\colon \mu_1 \neq \mu_2$

(b) d.f. $= n_1 + n_2 - 2 = 12 + 17 - 2 = 27$

$t_0 = \pm 1.703$; Reject H_0 if $t < -1.703$ or $t > 1.703$.

(c) $t = \dfrac{(\bar{x}_1 - \bar{x}_2) - (\mu_1 - \mu_2)}{\sqrt{\dfrac{(n_1 - 1)s_1^2 + (n_2 - 1)s_2^2}{n_1 + n_2 - 2}}\sqrt{\dfrac{1}{n_1} + \dfrac{1}{n_2}}}$

$= \dfrac{(10.1 - 8.3) - (0)}{\sqrt{\dfrac{(12 - 1)(4.11)^2 + (17 - 1)(4.02)^2}{12 + 17 - 2}}\sqrt{\dfrac{1}{12} + \dfrac{1}{17}}} = \dfrac{1.8}{\sqrt{16.459}\,\sqrt{0.142}} \approx 1.177$

(d) Fail to reject H_0.

(e) There is not enough evidence to reject the claim.

17. (a) $H_0\colon \mu_1 \geq \mu_2$; $H_a\colon \mu_1 < \mu_2$ (claim)

(b) d.f. $= n_1 + n_2 - 2 = 35$

$t_0 = -1.282$; Reject H_0 if $t < -1.282$.

(c) $t = \dfrac{(\bar{x}_1 - \bar{x}_2) - (\mu_1 - \mu_2)}{\sqrt{\dfrac{(n_1 - 1)s_1^2 + (n_2 - 1)s_2^2}{n_1 + n_2 - 2}}\sqrt{\dfrac{1}{n_1} + \dfrac{1}{n_2}}} = \dfrac{(473 - 741) - (0)}{\sqrt{\dfrac{(14 - 1)(190)^2 + (23 - 1)(205)^2}{14 + 23 - 2}}\sqrt{\dfrac{1}{14} + \dfrac{1}{23}}}$

$= \dfrac{-268}{\sqrt{39{,}824.286}\,\sqrt{0.115}} = -3.960$

(d) Reject H_0.

(e) There is enough evidence to support the claim.

19. (a) $H_0: \mu_1 \le \mu_2$; $H_a: \mu_1 > \mu_2$ (claim)

(b) d.f. $= \min\{n_1 - 1, n_2 - 1\} = 14$

$t_0 = 1.345$; Reject H_0 if $t > 1.345$.

(c) $z = \dfrac{(\bar{x}_1 - \bar{x}_2) - (\mu_1 - \mu_2)}{\sqrt{\dfrac{s_1^2}{n_1} + \dfrac{s_2^2}{n_2}}} = \dfrac{(42{,}200 - 37{,}900) - (0)}{\sqrt{\dfrac{(8600)^2}{19} + \dfrac{(5500)^2}{15}}} = \dfrac{4300}{\sqrt{5{,}909{,}298.246}} \approx 1.769$

(d) Reject H_0.

(e) There is enough evidence to support the claim.

21. (a) $H_0: \mu_1 = \mu_2$; $H_a: \mu_1 \ne \mu_2$ (claim)

(b) d.f. $= n_1 + n_2 - 2 = 21$

$t_0 \pm 2.831$; Reject H_0 if $t < -2.831$ or $t > 2.831$.

(c) $\bar{x}_1 = 340.300$, $s_1 = 22.301$, $n_1 = 10$

$\bar{x}_2 = 389.538$, $s_2 = 14.512$, $n_2 = 13$

$t = \dfrac{(\bar{x}_1 - \bar{x}_2) - (\mu_1 - \mu_2)}{\sqrt{\dfrac{(n_1 - 1)s_1^2 + (n_2 - 1)s_2^2}{n_1 + n_2 - 2}}\sqrt{\dfrac{1}{n_1} + \dfrac{1}{n_2}}}$

$= \dfrac{(340.300 - 389.538) - (0)}{\sqrt{\dfrac{(10 - 1)(22.301)^2 + (13 - 1)(14.512)^2}{10 + 13 - 2}}\sqrt{\dfrac{1}{10} + \dfrac{1}{13}}} = \dfrac{-49.238}{\sqrt{333.485}\,\sqrt{0.177}} \approx -6.410$

(d) Reject H_0.

(e) There is enough evidence to support the claim.

23. (a) $H_0: \mu_1 \ge \mu_2$; $H_a: \mu_1 < \mu_2$ (claim)

(b) d.f. $= n_1 + n_2 - 2 = 42$

$t_0 = -1.282 \rightarrow t < -1.282$

(c) $\bar{x}_1 = 56.684$, $s_1 = 6.961$, $n_1 = 19$

$\bar{x}_2 = 67.400$, $s_2 = 9.014$, $n_2 = 25$

$t = \dfrac{(\bar{x}_1 - \bar{x}_2) - (\mu_1 - \mu_2)}{\sqrt{\dfrac{(n_1 - 1)s_1^2 + (n_2 - 1)s_2^2}{n_1 + n_2 - 2}}\sqrt{\dfrac{1}{n_1} + \dfrac{1}{n_2}}}$

$= \dfrac{(56.684 - 67.400) - (0)}{\sqrt{\dfrac{(19 - 1)(6.961)^2 + (25 - 1)(9.014)^2}{19 + 25 - 2}}\sqrt{\dfrac{1}{19} + \dfrac{1}{25}}} = \dfrac{-10.716}{\sqrt{67.196}\,\sqrt{0.0926}} \approx -4.295$

(d) Reject H_0.

(e) There is enough evidence to support the claim and to recommend changing to the new method.

25. $\hat{\sigma} = \sqrt{\dfrac{(n_1 - 1)s_1^2 + (n_2 - 1)s_2^2}{n_1 + n_2 - 2}} = \sqrt{\dfrac{(15 - 1)(6.2)^2 + (12 - 1)(8.1)^2}{15 + 12 - 2}} \approx 7.099$

$(\bar{x}_1 - \bar{x}_2) \pm t_c \hat{\sigma} \sqrt{\dfrac{1}{n_1} + \dfrac{1}{n_2}} \rightarrow (450 - 420) \pm 2.060 \cdot 7.099 \sqrt{\dfrac{1}{15} + \dfrac{1}{12}}$

$\rightarrow 30 \pm 5.664 \rightarrow 24.336 < \mu_1 - \mu_2 < 35.664 \rightarrow 24 < \mu_1 - \mu_2 < 36$

27. $(\bar{x}_1 - \bar{x}_2) \pm t_c \sqrt{\dfrac{s_1^2}{n_1} + \dfrac{s_2^2}{n_2}} \rightarrow (75 - 70) \pm 1.771 \sqrt{\dfrac{(3.64)^2}{16} + \dfrac{(2.12)^2}{14}}$

$\rightarrow -5 \pm 1.898 \rightarrow 3.102 < \mu_1 - \mu_2 < 6.898 \rightarrow 3 < \mu_1 - \mu_2 < 7$

8.3 TESTING THE DIFFERENCE BETWEEN MEANS (DEPENDENT SAMPLES)

8.3 Try It Yourself Solutions

1.

Before	After	d	d^2
72	73	−1	1
81	80	1	1
76	79	−3	9
74	76	−2	4
75	76	−1	1
80	80	0	0
68	74	−6	36
75	77	−2	4
78	75	3	9
76	74	2	4
74	76	−2	4
77	78	−1	1
		$\Sigma d = -12$	$\Sigma d^2 = 74$

a. $H_0: \mu_d \geq 0$; $H_a: \mu_d < 0$ (claim)

b. $\alpha = 0.05$ and d.f. $= n - 1 = 11$

c. $t_0 \approx -1.796$; Reject H_0 if $t < -1.796$.

d. $\bar{d} = \dfrac{\Sigma d}{n} = \dfrac{-12}{12} = -1$

$s_d = \sqrt{\dfrac{n(\Sigma d^2) - (\Sigma d)^2}{n(n - 1)}} = \sqrt{\dfrac{12(74) - (-12)^2}{12(11)}} \approx 2.374$

e. $t = \dfrac{\bar{d} - \mu_d}{\dfrac{s_d}{\sqrt{n}}} = \dfrac{-1 - 0}{\dfrac{2.374}{\sqrt{12}}} \approx -1.459$

f. Fail to reject H_0.

g. There is not enough evidence to support the claim.

2.

Before	After	d	d^2
101.8	99.2	2.6	6.76
98.5	98.4	0.1	0.01
98.1	98.2	−0.1	0.01
99.4	99	0.4	0.16
98.9	98.6	0.3	0.09
100.2	99.7	0.5	0.25
97.9	97.8	0.1	0.01
		$\Sigma d = 3.9$	$\Sigma d^2 = 7.29$

a. $H_0: \mu_d = 0$; $H_a: \mu_d \neq 0$ (claim)

b. $\alpha = 0.05$ and d.f. $= n - 1 = 6$

c. $t_0 = \pm 2.447$; Reject H_0 if $t < -2.447$ or $t > 2.447$.

d. $\bar{d} = \dfrac{\Sigma d}{n} = \dfrac{3.9}{7} \approx 0.557$

$$s_d = \sqrt{\frac{n(\Sigma d^2) - (\Sigma d)^2}{n(n-1)}} = \sqrt{\frac{7(7.29) - (3.9)^2}{7(6)}} \approx 0.924$$

e. $t = \dfrac{\bar{d} - \mu_d}{\dfrac{s_d}{\sqrt{n}}} = \dfrac{0.557 - 0}{\dfrac{0.924}{\sqrt{7}}} \approx 1.595$

f. Fail to reject H_0.

g. There is not enough evidence at the 5% significance level to conclude that the drug changes the body's temperature.

8.3 EXERCISE SOLUTIONS

1. (1) Each sample must be randomly selected from a normal population.

   (2) Each member of the first sample must be paired with a member of the second sample.

3. $H_0: \mu_d \geq 0; H_a: \mu_d < 0$ (claim)

   $\alpha = 0.05$ and d.f. $= n - 1 = 13$

   $t_0 = -1.771$ (Left-tailed)

   $t = \dfrac{\bar{d} - \mu_d}{\dfrac{s_d}{\sqrt{n}}} = \dfrac{1.5 - 0}{\dfrac{3.2}{\sqrt{14}}} = \dfrac{1.5}{855} \approx 1.754$

   Fail to reject H_0.

5. $H_0: \mu_d \leq 0$ (claim); $H_a: \mu_d > 0$

   $\alpha = 0.10$ and d.f. $= n - 1 = 15$

   $t_0 = 1.341$ (Right-tailed)

   $t = \dfrac{\bar{d} - \mu_d}{\dfrac{s_d}{\sqrt{n}}} = \dfrac{6.5 - 0}{\dfrac{9.54}{\sqrt{16}}} = \dfrac{6.5}{2.385} \approx 2.725$

   Reject H_0.

7. $H_0: \mu_d \geq 0$ (claim); $H_a: \mu_d < 0$

   $\alpha = 0.01$ and d.f. $= n - 1 = 14$

   $t_0 = -2.624$ (Left-tailed)

   $t = \dfrac{\bar{d} - \mu_d}{\dfrac{s_d}{\sqrt{n}}} = \dfrac{-2.3 - 0}{\dfrac{1.2}{\sqrt{15}}} = \dfrac{-2.3}{0.3098} \approx -7.423$

   Reject H_0.

9. (a) $H_0: \mu_d \geq 0; H_a: \mu_d < 0$ (claim)

   (b) $t_0 = -2.650$; Reject H_0 if $t < -2.650$.

   (c) $\bar{d} \approx -33.714$ and $s_d \approx 42.034$

(d) $t = \dfrac{\bar{d} - \mu_d}{\dfrac{s_d}{\sqrt{n}}} = \dfrac{-33.714 - 0}{\dfrac{42.034}{\sqrt{14}}} = \dfrac{-33.714}{11.234} \approx -3.001$

(e) Reject H_0.

(f) There is enough evidence to support the claim that the SAT scores improved.

11. (a) $H_0: \mu_d \geq 0; H_a: \mu_d < 0$ (claim)

(b) $t_0 = -1.415$; Reject H_0 if $t < 1.415$.

(c) $\bar{d} \approx -1.575$ and $s_d \approx 0.803$

(d) $t = \dfrac{\bar{d} - \mu_d}{\dfrac{s_d}{\sqrt{n}}} = \dfrac{-1.575 - 0}{\dfrac{0.803}{\sqrt{8}}} = \dfrac{-1.575}{0.284} = -5.548$

(e) Reject H_0.

(f) There is enough evidence to support the fuel additive improved gas mileage.

13. (a) $H_0: \mu_d \leq 0; H_a: \mu > 0$ (claim)

(b) $t_0 = 1.363$; Reject H_0 if $t > 1.363$.

(c) $\bar{d} = 3.75$ and $s_d \approx 7.84$

(d) $t = \dfrac{\bar{d} - \mu_d}{\dfrac{s_d}{\sqrt{n}}} = \dfrac{3.75 - 0}{\dfrac{7.84}{\sqrt{12}}} = \dfrac{3.75}{2.26} = 1.657$

(e) Reject H_0.

(f) There is enough evidence to support the claim that the exercise program helps participants lose weight.

15. (a) $H_0: \mu_d \leq 0; H_a: \mu_d > 0$ (claim)

(b) $t_0 = 2.764$; Reject H_0 if $t > 2.764$.

(c) $\bar{d} \approx 1.255$ and $s_d \approx 0.4414$

(d) $t = \dfrac{\bar{d} - \mu_d}{\dfrac{s_d}{\sqrt{n}}} = \dfrac{1.255 - 0}{\dfrac{0.4414}{\sqrt{11}}} = \dfrac{1.255}{0.1331} \approx 9.438$

(e) Reject H_0.

(f) There is enough evidence to support the claim that soft tissue therapy and spinal manipulation help reduce the length of time patients suffer from headaches.

17. (a) $H_0: \mu_d \leq 0; H_a: \mu_d > 0$ (claim)

(b) $t_0 = 1.895$; Reject H_0 if $t > 1.895$.

(c) $\bar{d} = 14.75$ and $s_d \approx 6.861$

(d) $t = \dfrac{\bar{d} - \mu_d}{\dfrac{s_d}{\sqrt{n}}} = \dfrac{14.75 - 0}{\dfrac{6.86}{\sqrt{8}}} = \dfrac{14.75}{2.4257} \approx 6.081$

(e) Reject H_0.

(f) There is enough evidence to support the claim that the new drug reduces systolic blood pressure.

19. (a) $H_0: \mu_d = 0$; $H_a: \mu_d \neq 0$ (claim)

(b) $t_0 = \pm 2.365$; Reject H_0 if $t < -2.365$ or $t > 2.365$.

(c) $\bar{d} = -1$ and $s_d = 1.309$

(d) $t = \dfrac{\bar{d} - \mu_d}{\dfrac{s_d}{\sqrt{n}}} = \dfrac{-1 - 0}{\dfrac{1.31}{\sqrt{8}}} = \dfrac{-1}{0.463} = -2.160$

(e) Fail to reject H_0.

(f) There is not enough evidence to support the claim that the product ratings have changed.

21. $\bar{d} \approx -1.525$ and $s_d \approx 0.542$

$$\bar{d} - t_{\alpha/2}\frac{s_d}{\sqrt{n}} < \mu_d < \bar{d} - t_{\alpha/2}\frac{s_d}{\sqrt{n}}$$

$$-1.525 - 1.753\left(\frac{0.542}{\sqrt{16}}\right) < \mu_d < -1.525 + 1.753\left(\frac{0.542}{\sqrt{16}}\right)$$

$$-1.525 - 0.238 < \mu_d < -1.525 + 0.238$$

$$-1.76 < \mu_d < -1.29$$

8.4 TESTING THE DIFFERENCE BETWEEN PROPORTIONS

8.4 Try It Yourself Solutions

1a. $H_0: p_1 = p_2$; $H_a: p_1 \neq p_2$ (claim)   **b.** $\alpha = 0.05$

c. $z_0 = \pm 1.96$; Reject H_0 if $z < -1.96$ or $z > 1.96$.

d. $\bar{p} = \dfrac{x_1 + x_2}{n_1 + n_2} = \dfrac{1484 + 1497}{6869 + 6869} = 0.217$

$\bar{q} = 0.783$

e. $n_1\bar{p} \approx 1490.573 > 5$, $n_1\bar{q} \approx 5378.427 > 5$, $n_2\bar{p} \approx 1490.573 > 5$, and $n_2\bar{q} \approx 5378.427 > 5$.

f. $z = \dfrac{(\hat{p}_1 - \hat{p}_2) - (p_1 - p_2)}{\sqrt{\bar{p}\,\bar{q}\left(\dfrac{1}{n_1} + \dfrac{1}{n_2}\right)}} = \dfrac{(0.216 - 0.218) - (0)}{\sqrt{0.217 \cdot 0.783\left(\dfrac{1}{6869} + \dfrac{1}{6869}\right)}} = \dfrac{-0.002}{\sqrt{0.00004947}} \approx -0.284$

g. Fail to reject H_0.

h. There is not enough evidence to support the claim.

2a. $H_0: p_1 \leq p_2$; $H_a: p_1 > p_2$ (claim)   **b.** $\alpha = 0.05$

c. $z_0 = 1.645$; Reject H_0 if $z > 1.645$.

d. $\bar{p} = \dfrac{x_1 + x_2}{n_1 + n_2} = \dfrac{1264 + 522}{6869 + 6869} = 0.130$

$\bar{q} = 0.870$

e. $n_1 \bar{p} \approx 892.97 > 5$, $n_1 \bar{q} \approx 5976.03 > 5$, $n_2 \bar{p} \approx 892.97 > 5$, and $n_2 \bar{q} \approx 5976.03 > 5$.

f. $z = \dfrac{(\hat{p}_1 - \hat{p}_2) - (p_1 - p_2)}{\sqrt{\bar{p}\,\bar{q}\left(\dfrac{1}{n_1} + \dfrac{1}{n_2}\right)}} = \dfrac{(0.184 - 0.076) - (0)}{\sqrt{0.130 \cdot 0.870\left(\dfrac{1}{6869} + \dfrac{1}{6869}\right)}} = \dfrac{0.108}{\sqrt{0.00003293}} \approx 18.820$

g. Reject H_0.

h. There is enough evidence to support the claim.

8.4 EXERCISE SOLUTIONS

1. State the hypotheses and identify the claim. Specify the level of significance. Find the critical value(s) and rejection region(s). Find $\bar{p}$ and $\bar{q}$. Find the standardized test statistic. Make a decision and interpret in the context of the claim.

3. $H_0: p_1 = p_2$; $H_a: p_1 \neq p_2$ (claim)

$z_0 = \pm 2.575$ (Two-tailed test)

$\bar{p} = \dfrac{x_1 + x_2}{n_1 + n_2} = \dfrac{35 + 36}{70 + 60} = 0.546$

$\bar{q} = 0.454$

$z = \dfrac{(\hat{p}_1 - \hat{p}_2) - (p_1 - p_2)}{\sqrt{\bar{p}\,\bar{q}\left(\dfrac{1}{n_1} + \dfrac{1}{n_2}\right)}} = \dfrac{(0.500 - 0.600) - (0)}{\sqrt{0.546 \cdot 0.454\left(\dfrac{1}{70} + \dfrac{1}{60}\right)}} = \dfrac{-0.100}{\sqrt{0.00767}} \approx -1.142$

Fail to reject H_0.

5. $H_0: p_1 \leq p_2$ (claim); $H_a: p_1 > p_2$

$z_0 = 1.282$ (Right-tailed test)

$\bar{p} = \dfrac{x_1 + x_2}{n_1 + n_2} = \dfrac{344 + 304}{860 + 800} = 0.390$

$\bar{q} = 0.610$

$z = \dfrac{(\hat{p}_1 - \hat{p}_2) - (p_1 - p_2)}{\sqrt{\bar{p}\,\bar{q}\left(\dfrac{1}{n_1} + \dfrac{1}{n_2}\right)}} = \dfrac{(0.400 - 0.380) - (0)}{\sqrt{0.390 \cdot 0.610\left(\dfrac{1}{860} + \dfrac{1}{800}\right)}} = \dfrac{0.020}{\sqrt{0.0000574003}} \approx 0.835$

Fail to reject H_0.

7. (a) $H_0: p_1 = p_2$ (claim); $H_a: p_1 \neq p_2$

(b) $z_0 = \pm 1.96$; Reject H_0 if $z < -1.96$ or $z > 1.96$.

$\bar{p} = \dfrac{x_1 + x_2}{n_1 + n_2} = \dfrac{520 + 865}{1539 + 2055} = 0.385$

$\bar{q} = 0.615$

(c) $z = \dfrac{(\hat{p}_1 - \hat{p}_2) - (p_1 - p_2)}{\sqrt{\bar{p}\,\bar{q}\left(\dfrac{1}{n_1} + \dfrac{1}{n_2}\right)}} = \dfrac{(0.338 - 0.421) - (0)}{\sqrt{0.385 \cdot 0.615\left(\dfrac{1}{1539} + \dfrac{1}{2055}\right)}}$

$= \dfrac{-0.083}{\sqrt{0.0000269069}} \approx -5.06$

(d) Reject H_0.

(e) There is sufficient evidence at the 5% level to reject the claim that the proportion of adults using alternative medicines has not changed since 1991.

9. (a) $H_0: p_1 = p_2$ (claim); $H_a: p_1 \neq p_2$

(b) $z_0 = \pm 1.645$; Reject H_0 if $z < -1.645$ or $z > 1.645$.

$$\bar{p} = \frac{x_1 + x_2}{n_1 + n_2} = \frac{2201 + 2348}{5240 + 6180} = 0.398$$

$$\bar{q} = 0.602$$

(c) $z = \dfrac{(\hat{p}_1 - \hat{p}_2) - (p_1 - p_2)}{\sqrt{\bar{p}\,\bar{q}\left(\dfrac{1}{n_1} + \dfrac{1}{n_2}\right)}} = \dfrac{(0.4200 - 0.37994) - 0}{\sqrt{(0.398)(0.602)\left(\dfrac{1}{5240} + \dfrac{1}{6180}\right)}}$

$$= \frac{0.041}{\sqrt{0.000084494}} = 4.362$$

(d) Reject H_0.

(e) There is sufficient evidence at the 10% significance level to reject the claim that the proportions of male and female senior citizens that eat the daily recommended number of servings of vegetables are the same.

11. (a) $H_0: p_1 \leq p_2$; $H_a: p_1 > p_2$ (claim)

(b) $z_0 = 2.33$; Reject H_0 if $z > 2.33$.

$$\bar{p} = \frac{x_1 + x_2}{n_1 + n_2} = \frac{496 + 468}{2000 + 2000} = 0.241$$

$$\bar{q} = 0.759$$

(c) $z = \dfrac{(\hat{p}_1 - \hat{p}_2) - (p_1 - p_2)}{\sqrt{\bar{p}\,\bar{q}\left(\dfrac{1}{n_1} + \dfrac{1}{n_2}\right)}} = \dfrac{(0.248 - 0.234) - (0)}{\sqrt{0.241 \cdot 0.759\left(\dfrac{1}{2000} + \dfrac{1}{2000}\right)}} = \dfrac{0.014}{\sqrt{0.0001829}} \approx 1.04$

(d) Fail to reject H_0.

(e) There is not sufficient evidence at the 1% significance level to support the claim that the proportion of adults who are smokers is greater in Alabama than in Missouri.

13. (a) $H_0: p_1 \geq p_2$; $H_a: p_1 < p_2$ (claim)

(b) $z_0 = -2.33$; Reject H_0 if $z < -2.33$.

$$\bar{p} = \frac{x_1 + x_2}{n_1 + n_2} = \frac{2083 + 985}{9300 + 4900} = 0.216$$

$$\bar{q} = 0.784$$

(c) $z = \dfrac{(\hat{p}_1 - \hat{p}_2) - (p_1 - p_2)}{\sqrt{\bar{p}\,\bar{q}\left(\dfrac{1}{n_1} + \dfrac{1}{n_2}\right)}} = \dfrac{(0.22398 - 0.20102) - (0)}{\sqrt{0.216 \cdot 0.784\left(\dfrac{1}{9300} + \dfrac{1}{4900}\right)}} = \dfrac{0.02296}{\sqrt{0.00005277}} \approx 3.16$

(d) Fail to reject H_0.

(e) There is sufficient evidence at the 1% significance level to support the claim that the proportion of twelfth grade males who said they had smoked in the last 30 days is less than the proportion of twelfth grade females.

15. (a) $H_0: p_1 = p_2$ (claim); $H_a: p_1 \neq p_2$

    (b) $z_0 = \pm 1.96$; Reject H_0 if $z < -1.96$ or $z > 1.96$.

$$\bar{p} = \frac{x_1 + x_2}{n_1 + n_2} = \frac{805 + 746}{1150 + 1050} = 0.705$$

$$\bar{q} = 0.295$$

    (c) $z = \dfrac{(\hat{p}_1 - \hat{p}_2) - (p_1 - p_2)}{\sqrt{\bar{p}\,\bar{q}\left(\dfrac{1}{n_1} + \dfrac{1}{n_2}\right)}} = \dfrac{(0.70000 - 0.71047) - 0}{\sqrt{(0.705)(0.295)\left(\dfrac{1}{1150} + \dfrac{1}{1050}\right)}} = \dfrac{-0.01047}{\sqrt{0.0003789}} = -0.54$

    (d) Fail to reject H_0.

    (e) There is insufficient evidence at the 5% significance level to reject the claim that the proportions of Internet users are the same for both groups.

17. $H_0: p_1 \geq p_2$; $H_a: p_1 < p_2$ (claim)

$$z_0 = -2.33$$

$$\bar{p} = \frac{x_1 + x_2}{n_1 + n_2} = \frac{28 + 35}{700 + 500} = 0.053$$

$$\bar{q} = 0.947$$

$$z = \frac{(\hat{p}_1 - \hat{p}_2) - (p_1 - p_2)}{\sqrt{\bar{p}\,\bar{q}\left(\dfrac{1}{n_1} + \dfrac{1}{n_2}\right)}} = \frac{(0.04 - 0.07) - (0)}{\sqrt{0.053 \cdot 0.947\left(\dfrac{1}{700} + \dfrac{1}{500}\right)}} = \frac{-0.03}{\sqrt{0.0001721}} \approx -2.287$$

Fail to reject H_0. There is insufficient evidence at the 1% significance level to support the claim.

19. $H_0: p_1 \geq p_2$; $H_a: p_1 < p_2$ (claim)

$$z_0 = -1.645$$

$$\bar{p} = \frac{x_1 + x_2}{n_1 + n_2} = \frac{189 + 185}{700 + 500} = 0.312$$

$$\bar{q} = 0.688$$

$$z = \frac{(\hat{p}_1 - \hat{p}_2) - (p_1 - p_2)}{\sqrt{\bar{p}\,\bar{q}\left(\dfrac{1}{n_1} + \dfrac{1}{n_2}\right)}} = \frac{(0.27 - 0.37) - (0)}{\sqrt{0.312 \cdot 0.688\left(\dfrac{1}{700} + \dfrac{1}{500}\right)}} = \frac{-0.10}{\sqrt{0.0007359}} \approx -3.686$$

Reject H_0. There is sufficient evidence at the 5% significance level to support the organization's claim.

21. $H_0: p_1 \leq p_2$; $H_a: p_1 > p_2$ (claim)

$$z_0 = 1.645$$

$$\bar{p} = \frac{x_1 + x_2}{n_1 + n_2} = \frac{7501 + 7501}{13,300 + 14,100} = 0.548$$

$$\bar{q} = 0.452$$

$$z = \frac{(\hat{p}_1 - \hat{p}_2) - (p_1 - p_2)}{\sqrt{\bar{p}\,\bar{q}\left(\dfrac{1}{n_1} + \dfrac{1}{n_2}\right)}} = \frac{(0.564 - 0.532) - (0)}{\sqrt{(0.548)(0.452)\left(\dfrac{1}{13,300} + \dfrac{1}{14,100}\right)}} = \frac{0.032}{\sqrt{0.00003619}} \approx 5.319$$

Reject H_0. There is sufficient evidence at the 5% significance level to support the claim.

23. $H_0: p_1 = p_2$ (claim); $H_a: p_1 \neq p_2$

$z_0 = \pm 2.576$

$\bar{p} = \dfrac{x_1 + x_2}{n_1 + n_2} = \dfrac{7501 + 5610}{13,300 + 13,200} = 0.495$

$\bar{q} = 0.505$

$z = \dfrac{(\hat{p}_1 - \hat{p}_2) - (p_1 - p_2)}{\sqrt{\bar{p}\bar{q}\left(\dfrac{1}{n_1} + \dfrac{1}{n_2}\right)}} = \dfrac{(0.564 - 0.425) - (0)}{\sqrt{(0.495)(0.505)\left(\dfrac{1}{13,300} + \dfrac{1}{13,200}\right)}} = \dfrac{0.139}{\sqrt{0.00003773}} \approx 22,629$

Reject H_0. There is sufficient evidence at the 1% significance level to reject the claim.

25. $(\hat{p}_1 - \hat{p}_2) \pm z_c\sqrt{\dfrac{\hat{p}_1\hat{q}_1}{n_1} + \dfrac{\hat{p}_2\hat{q}_2}{n_2}} \rightarrow (0.088 - 0.083) \pm 1.96\sqrt{\dfrac{0.088 \cdot 0.912}{1,068,000} + \dfrac{0.083 \cdot 0.917}{1,476,000}}$

$\rightarrow 0.005 \pm 1.96\sqrt{0.00000012671}$

$\rightarrow 0.004 < p_1 - p_2 < 0.006$

CHAPTER 8 REVIEW EXERCISE SOLUTIONS

1. Independent because the two samples of laboratory mice are different.

3. $H_0: \mu_1 \geq \mu_2$ (claim); $H_1: \mu_1 < \mu_2$

$z_0 = -1.645$

$z = \dfrac{(\bar{x}_1 - \bar{x}_2) - (\mu_1 - \mu_2)}{\sqrt{\dfrac{s_1^2}{n_1} + \dfrac{s_2^2}{n_2}}} = \dfrac{(1.28 - 1.34) - (0)}{\sqrt{\dfrac{(0.30)^2}{96} + \dfrac{(0.23)^2}{85}}} = \dfrac{-0.06}{\sqrt{0.001560}} \approx -1.519$

Fail to reject H_0. There is not enough evidence to reject the claim.

5. $H_0: \mu_1 \geq \mu_2$; $H_1: \mu_1 < \mu_2$ (claim)

$z_0 = -1.282$

$z = \dfrac{(\bar{x}_1 - \bar{x}_2) - (\mu_1 - \mu_2)}{\sqrt{\dfrac{s_1^2}{n_1} + \dfrac{s_2^2}{n_2}}} = \dfrac{(0.28 - 0.33) - (0)}{\sqrt{\dfrac{(0.11)^2}{41} + \dfrac{(0.10)^2}{34}}} = \dfrac{-0.50}{\sqrt{0.00058924}} \approx -2.060$

Reject H_0. There is enough evidence to support the claim.

7. (a) $H_0: \mu_1 \leq \mu_2$; $H_1: \mu_1 > \mu_2$ (claim)

(b) $z_0 = 1.645$; Reject H_0 if $z > 1.645$.

(c) $z = \dfrac{(\bar{x}_1 - \bar{x}_2) - (\mu_1 - \mu_2)}{\sqrt{\dfrac{s_1^2}{n_1} + \dfrac{s_2^2}{n_2}}} = \dfrac{(480 - 470) - (0)}{\sqrt{\dfrac{(32)^2}{36} + \dfrac{(54)^2}{41}}} = \dfrac{10}{\sqrt{99.566}} \approx 1.002$

(d) Fail to reject H_0.

(e) There is not enough evidence to support the claim.

9. H_0: $\mu_1 = \mu_2$ (claim); H_a: $\mu_1 \neq \mu_2$

d.f. $= n_1 + n_2 - 2 = 29$

$t_0 = \pm 2.045$

$$t = \frac{(\bar{x}_1 - \bar{x}_2) - (\mu_1 - \mu_2)}{\sqrt{\dfrac{(n_1-1)s_1^2 + (n_2-1)s_2^2}{n_1+n_2-2}}\sqrt{\dfrac{1}{n_1}+\dfrac{1}{n_2}}} = \frac{(300-290)-(0)}{\sqrt{\dfrac{(19-1)(26)^2+(12-1)(22)^2}{19+12-2}}\sqrt{\dfrac{1}{19}+\dfrac{1}{12}}}$$

$$= \frac{10}{\sqrt{603.172}\,\sqrt{0.1360}} \approx 1.104$$

Fail to reject H_0. There is not enough evidence to reject the claim.

11. H_0: $\mu_1 \leq \mu_2$ (claim); H_a: $\mu_1 > \mu_2$

d.f. $= \min\{n_1 - 1, n_2 - 1\} = 24$

$t_0 = 1.711$

$$t = \frac{(\bar{x}_1 - \bar{x}_2) - (\mu_1 - \mu_2)}{\sqrt{\dfrac{s_1^2}{n_1}+\dfrac{s_2^2}{n_2}}} = \frac{(183.5-184.7)-(0)}{\sqrt{\dfrac{(1.3)^2}{25}+\dfrac{(3.9)^2}{25}}} = \frac{-1.2}{\sqrt{0.676}} \approx -1.460$$

Fail to reject H_0. There is not enough evidence to reject the claim.

13. H_0: $\mu_1 = \mu_2$; H_a: $\mu_1 \neq \mu_2$ (claim)

d.f. $= n_1 + n_2 - 2 = 10$

$t_0 = \pm 3.169$

$$t = \frac{(\bar{x}_1 - \bar{x}_2) - (\mu_1 - \mu_2)}{\sqrt{\dfrac{(n_1-1)s_1^2 + (n_2-1)s_2^2}{n_1+n_2-2}} \cdot \sqrt{\dfrac{1}{n_1}+\dfrac{1}{n_2}}} = \frac{(61-55)-(0)}{\sqrt{\dfrac{(5-1)\,3.3^2+(7-1)\,1.2^2}{5+7-2}} \cdot \sqrt{\dfrac{1}{5}+\dfrac{1}{7}}}$$

$$= \frac{6}{\sqrt{5.22}\,\sqrt{0.343}} \approx 4.484$$

Reject H_0. There is enough evidence to support the claim.

15. (a) H_0: $\mu_1 \leq \mu_2$; H_a: $\mu_1 > \mu_2$ (claim)

 (b) d.f. $= n_1 + n_2 - 2 = 42$

 $t_0 = 1.645$; Reject H_0 if $t > 1.645$.

 (c) $\bar{x}_1 = 51.476$, $s_1 = 11.007$, $n_1 = 21$

 $\bar{x}_2 = 41.522$, $s_2 = 17.149$, $n_2 = 23$

$$t = \frac{(\bar{x}_1 - \bar{x}_2) - (\mu_1 - \mu_2)}{\sqrt{\dfrac{(n_1-1)s_1^2 + (n_2-1)s_2^2}{n_1+n_2-2}}\sqrt{\dfrac{1}{n_1}+\dfrac{1}{n_2}}} = \frac{(51.476-41.522)-(0)}{\sqrt{\dfrac{(21-1)(11.007)^2+(23-1)(17.149)^2}{21+23-2}}\sqrt{\dfrac{1}{21}+\dfrac{1}{23}}}$$

$$= \frac{9.954}{\sqrt{211.7386}\,\sqrt{0.0911}} \approx 2.266$$

 (d) Reject H_0.

 (e) There is sufficient evidence at the 5% significance level to support the claim that the third graders taught with the directed reading activities scored higher than those taught without the activities.

17. $H_0: \mu_d = 0$ (claim); $H_a: \mu_d \neq 0$

$\alpha = 0.05$ and d.f. $= n - 1 = 99$

$t_0 = \pm 1.96$ (Two-tailed test)

$t = \dfrac{\bar{d} - \mu_d}{\dfrac{s_d}{\sqrt{n}}} = \dfrac{10 - 0}{\dfrac{12.4}{\sqrt{100}}} = \dfrac{10}{1.24} \approx 8.065$

Reject H_0.

19. $H_0: \mu_d \leq 6$ (claim); $H_a: \mu_d > 6$

$\alpha = 0.10$ and d.f. $= n - 1 = 32$

$t_0 = 1.282$ (Right-tailed test)

$t = \dfrac{\bar{d} - \mu_d}{\dfrac{s_d}{\sqrt{n}}} = \dfrac{10.3 - 6}{\dfrac{1.24}{\sqrt{33}}} = \dfrac{4.3}{0.21586} \approx 19.921$

Reject H_0.

21. (a) $H_0: \mu_d \leq 0; H_a: \mu_d > 0$ (claim)

(b) $t_0 = 1.383$; Reject H_0 if $t > 1.383$.

(c) $\bar{d} = 5$ and $s_d \approx 8.743$

(d) $t = \dfrac{\bar{d} - \mu_d}{\dfrac{s_d}{\sqrt{n}}} = \dfrac{5 - 0}{\dfrac{8.743}{\sqrt{10}}} = \dfrac{5}{2.765} \approx 1.808$

(e) Reject H_0.

(f) There is enough evidence to support the claim.

23. $H_0: p_1 = p_2$; $H_a: p_1 \neq p_2$ (claim)

$z_0 = \pm 1.96$ (Two-tailed test)

$\bar{p} = \dfrac{x_1 + x_2}{n_1 + n_2} = \dfrac{425 + 410}{840 + 760} = 0.522$

$\bar{q} = 0.478$

$z = \dfrac{(\hat{p}_1 - \hat{p}_2) - (p_1 - p_2)}{\sqrt{\bar{p}\bar{q}\left(\dfrac{1}{n_1} + \dfrac{1}{n_2}\right)}} = \dfrac{(0.506 - 0.539) - (0)}{\sqrt{0.522 \cdot 0.478\left(\dfrac{1}{840} + \dfrac{1}{760}\right)}} = \dfrac{-0.033}{\sqrt{0.0006254}} \approx -1.320$

Fail to reject H_0.

25. $H_0: p_1 \leq p_2; H_a: p_1 > p_2$ (claim)

$z_0 = 1.282$ (Right-tailed test)

$$\bar{p} = \frac{x_1 + x_2}{n_1 + n_2} = \frac{261 + 207}{556 + 483} = 0.450$$

$\bar{q} = 0.550$

$$z = \frac{(\hat{p}_1 - \hat{p}_2) - (p_1 - p_2)}{\sqrt{\bar{p}\,\bar{q}\left(\frac{1}{n_1} + \frac{1}{n_2}\right)}} = \frac{(0.469 - 0.429) - (0)}{\sqrt{0.450 \cdot 0.550\left(\frac{1}{556} + \frac{1}{483}\right)}} = \frac{0.04}{\sqrt{0.0009576}} \approx 1.293$$

Reject H_0.

27. (a) $H_0: p_1 = p_2$ (claim); $H_a: p_1 \neq p_2$

(b) $z_0 = \pm 1.645$; Reject H_0 if $z < -1.645$ or $z > 1.645$.

$$\bar{p} = \frac{x_1 + x_2}{n_1 + n_2} = \frac{398 + 530}{800 + 1000} = 0.516$$

$\bar{q} = 0.484$

(c) $$z = \frac{(\hat{p}_1 - \hat{p}_2) - (p_1 - p_2)}{\sqrt{\bar{p}\,\bar{q}\left(\frac{1}{n_1} + \frac{1}{n_2}\right)}} = \frac{(0.4975 - 0.5300) - (0)}{\sqrt{0.516 \cdot 0.484\left(\frac{1}{800} + \frac{1}{1000}\right)}} = \frac{-0.0325}{\sqrt{0.0005619}} \approx -1.371$$

(d) Fail to reject H_0.

(e) There is not enough evidence to reject the claim.

CHAPTER 8 QUIZ SOLUTIONS

1. (a) $H_0: \mu_1 \leq \mu_2; H_a: \mu_1 > \mu_2$ (claim)

(b) n_1 and $n_2 > 30$ and the samples are independent $\rightarrow$ Right tailed z-test

(c) $z_0 = 1.645$; Reject H_0 if $z > 1.645$.

(d) $$z = \frac{(\bar{x}_1 - \bar{x}_2) - (\mu_1 - \mu_2)}{\sqrt{\frac{s_1^2}{n_1} + \frac{s_2^2}{n_2}}} = \frac{(149 - 145) - (0)}{\sqrt{\frac{(35)^2}{49} + \frac{(33)^2}{50}}} = \frac{4}{\sqrt{46.780}} \approx 0.585$$

(e) Fail to reject H_0.

(f) There is not enough evidence at the 5% significance level to support the claim that the mean score on the science assessment for male high school students was higher than for the female high school students.

2. (a) $H_0: \mu_1 = \mu_2$ (claim); $H_a: \mu_1 \neq \mu_2$

(b) n_1 and $n_2 < 30$, the samples are independent, and the populations are normally distributed. $\rightarrow$ Two-tailed t-test (assume variances are equal)

(c) d.f. $= n_1 + n_2 - 2 = 26$

$t_0 = \pm 2.779$; Reject H_0 if $t < -2.779$ or $t > 2.779$.

(d) $t = \dfrac{(\bar{x}_1 - \bar{x}_2) - (\mu_1 - \mu_2)}{\sqrt{\dfrac{(n_1 - 1)s_1^2 + (n_2 - 1)s_2^2}{n_1 + n_2 - 2}} \sqrt{\dfrac{1}{n_1} + \dfrac{1}{n_2}}} = \dfrac{(153 - 149) - (0)}{\sqrt{\dfrac{(13 - 1)(32)^2 + (15 - 1)(30)^2}{13 + 15 - 2}} \sqrt{\dfrac{1}{13} + \dfrac{1}{15}}}$

$= \dfrac{4.0}{\sqrt{957.23}\,\sqrt{0.14359}} \approx 0.341$

(e) Fail to reject H_0.

(f) There is not enough evidence at the 1% significance level to reject the teacher's claim that the mean scores on the science assessment test are the same for fourth grade boys and girls.

3. (a) $H_0: p_1 \leq p_2;\ H_a: p_1 > p_2$ (claim)

(b) Testing 2 proportions, $n_1\,\bar{p},\ n_1\,\bar{q},\ n_2\,\bar{p}$, and $n_2\,\bar{q} \geq 5$, and the samples are independent $\rightarrow$ Right-tailed z-test

(c) $z_0 = 1.28$; Reject H_0 if $z > 1.28$.

(d) $\bar{p} = \dfrac{x_1 + x_2}{n_1 + n_2} = \dfrac{2043 + 3018}{6382 + 11{,}179} = 0.288$

$\bar{q} = 0.712$

$z = \dfrac{(\hat{p}_1 - \hat{p}_2) - (p_1 - p_2)}{\sqrt{\bar{p}\,\bar{q}\left(\dfrac{1}{n_1} + \dfrac{1}{n_2}\right)}} = \dfrac{(0.32 - 0.27) - (0)}{\sqrt{0.288 \cdot 0.712\left(\dfrac{1}{6382} + \dfrac{1}{11{,}179}\right)}}$

$= \dfrac{0.05}{\sqrt{0.000050473}} \approx 7.04$

(e) Reject H_0.

(f) There is sufficient evidence at the 10% significance level to support the claim that the proportion of fatal crashes involving alcohol is higher for drivers in the 21 to 24 age group than for drivers ages 25 to 35.

4. (a) $H_0: \mu_d \geq 0;\ H_a: \mu_d < 0$ (claim)

(b) Dependent samples and both populations are normally distributed. $\rightarrow$ one-tailed t-test

(c) $t_0 = -1.796$; Reject H_0 if $t < -1.796$.

(d) $t = \dfrac{\bar{d} - \mu_d}{\dfrac{s_d}{\sqrt{n}}} = \dfrac{0 - 68.5}{\dfrac{26.318}{\sqrt{12}}} = \dfrac{-68.5}{7.597} \approx -9.016$

(e) Reject H_0.

(f) There is sufficient evidence at the 5% significance level to conclude that the students' SAT scores improved on the second test.

1. (a) $\hat{p} = \dfrac{x}{n} = \dfrac{570}{1000} = 0.570$

$\hat{q} = 0.430$

$n\hat{p} = 570 \geq 5$

$n\hat{q} = 430 \geq 5$

Use normal distribution

$\hat{p} \pm z_c \sqrt{\dfrac{\hat{p}\hat{q}}{n}} = 0.570 \pm 1.96 \sqrt{\dfrac{(0.570)(0.430)}{1000}} = 0.570 \pm 0.031 \Rightarrow (0.539, 0.601)$

(b) Given that the 95% CI is (0.539, 0.601), it is unlikely that more than 60% of adults believe it is somewhat or very likely that life exists on other planets.

2. H_0: $\mu_d \leq 0$

H_a: $\mu_d > 0$ (claim)

$\bar{d} = 6.833$

$s_d = 3.713$

d.f. $= n - 1 = 11$

$t_0 = 1.796$

$t = \dfrac{\bar{d} - 0}{\dfrac{sd}{\sqrt{n}}} = \dfrac{6.833 - 0}{\dfrac{3.713}{\sqrt{12}}} = 6.375$

Reject H_0. There is enough evidence to support the claim.

3. $\bar{x} \pm z_c \dfrac{s}{\sqrt{n}} = 29.97 \pm 1.96 \dfrac{3.4}{\sqrt{42}} = 26.97 \pm 1.03 \Rightarrow (25.94, 28.00)$; z-distribution

4. $\bar{x} \pm t_c \dfrac{s}{\sqrt{n}} = 3.46 \pm 1.753 \dfrac{1.63}{\sqrt{16}} = 3.46 \pm 0.71 \Rightarrow (2.75, 4.17)$; t-distribution

5. $\bar{x} \pm z_c \dfrac{s}{\sqrt{n}} = 12.1 \pm 2.787 \dfrac{2.64}{\sqrt{26}} = 12.1 \pm 1.4 \Rightarrow (10.7, 13.5)$; t-distribution

6. $\bar{x} \pm t_c \dfrac{s}{\sqrt{n}} = 8.21 \pm 2.365 \dfrac{0.62}{\sqrt{8}} = 8.21 \pm 0.52 \Rightarrow (7.69, 8.73)$; t-distribution

7. H_0: $\mu_1 \leq \mu_2$

H_a: $\mu_1 > \mu_2$ (claim)

$z = 1.282$

$z = \dfrac{(\bar{x}_1 - \bar{x}_2) - 0}{\sqrt{\dfrac{s_1^2}{n_1} + \dfrac{s_2^2}{n_2}}} = \dfrac{3086 - 2263}{\sqrt{\dfrac{(563)^2}{85} + \dfrac{(624)^2}{68}}} = \dfrac{823}{\sqrt{9790.282}} = 8.318$

Reject H_0. There is enough evidence to support the claim.

8. H_0: $\mu \geq 33$

H_a: $\mu < 33$ (claim)

9. H_0: $p \geq 0.19$ (claim)

H_a: $p < 0.19$

10. H_0: $\sigma = 0.63$ (claim)

H_a: $\sigma \neq 0.63$

11. H_0: $\mu = 2.28$

H_a: $\mu \neq 2.28$ (claim)

12. (a) $\dfrac{(n-1)s^2}{\chi_R^2} < \sigma^2 < \dfrac{(n-1)s^2}{\chi_L^2} \Rightarrow \dfrac{(26-1)(3.1)^2}{46.928} < \sigma^2 < \dfrac{(26-1)(3.1)^2}{10.520}$

$\Rightarrow 5.1 < \sigma < 22.8$

(b) $\dfrac{(n-1)s^2}{\chi_R^2} < \sigma < \dfrac{(n-1)s^2}{\chi_L^2} \Rightarrow \sqrt{5.1} < \sigma < \sqrt{22.8}$

$\Rightarrow 2.3 < \sigma < 4.8$

(c) Because the 99% CI of σ is (2.3, 4.8) and contains 2.5, there is not enough evidence to support the pharmacist's claim.

13. H_0: $\mu_1 \geq \mu_2$

H_a: $\mu_1 < \mu_2$ (claim)

d.f. $= n_1 + n_2 - 2 = 15 + 15 - 2 = 28$

$t_0 = 1.701$

$t = \dfrac{(\bar{x}_1 - \bar{x}_2) - 0}{\sqrt{\dfrac{(n_1-1)s_1^2 + (n_2-1)s_2^2}{n_1 + n_2 - 2}} \sqrt{\dfrac{1}{n_1} + \dfrac{1}{n_2}}}$

$= \dfrac{57.9 - 61.1}{\sqrt{\dfrac{(15-1)(0.8)^2 + (15-1)(0.6)^2}{15 + 15 - 2}} \sqrt{\dfrac{1}{15} + \dfrac{1}{15}}}$

$= \dfrac{-3.2}{\sqrt{0.5}\sqrt{0.133}} = -12.394$

Reject H_0. There is enough evidence to support the claim.

14. (a) $\bar{x} = 296.231$

$s = 111.533$

$\bar{x} + t_c \dfrac{s}{\sqrt{n}} \Rightarrow 296.231 \pm 2.060 \dfrac{111.533}{\sqrt{26}}$

$\Rightarrow 296.231 \pm 45.059$

$\Rightarrow (251.2, 341.3)$

(b) H_0: $\mu \geq 280$

H_a: $\mu < 280$ (claim)

$t_0 = -1.316$

$t = \dfrac{\bar{x} - 280}{\dfrac{s}{\sqrt{n}}} = \dfrac{296.231 - 280}{\dfrac{111.533}{\sqrt{26}}} = 0.742$

Fail to reject H_0. There is not enough evidence to support the claim.

15. (a) H_0: $p_1 = p_2$ (claim)

H_a: $p_1 \neq p_2$

$\bar{p} = \dfrac{x_1 + x_2}{n_1 + n_2} = \dfrac{195 + 204}{319 + 323} = 0.621$

$\bar{q} = 0.379$

$z_0 = \pm 1.645$

$z = \dfrac{(\hat{p}_1 - \hat{p}_2) - 0}{\sqrt{\bar{p}\,\bar{q}\left(\dfrac{1}{n_1} + \dfrac{1}{n_2}\right)}}$

$= \dfrac{(0.611 - 0.632)}{\sqrt{(0.621)(0.379)\left(\dfrac{1}{319} + \dfrac{1}{323}\right)}}$

$= \dfrac{-0.021}{\sqrt{0.001466}} = -0.548$

Fail to reject H_0. There is not enough evidence to reject the claim.

16. $\bar{x} \pm z\dfrac{s}{\sqrt{n}} \Rightarrow 150 \pm 1.645\dfrac{34}{\sqrt{120}} \Rightarrow 150 \pm 5 \Rightarrow (145, 155)$

(b) H_0: $\mu \geq 145$ (claim)

H_a: $\mu < 145$

$z_0 = -1.645$

$z = \dfrac{\bar{x} - 145}{\dfrac{s}{\sqrt{n}}} = \dfrac{150 - 145}{\dfrac{34}{\sqrt{120}}} = 1.611$

Fail to reject H_0. There is not enough evidence to reject the claim.

9.1 Try It Yourself Solutions

1ab.

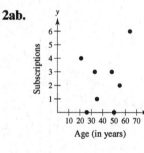

Donating percent
Family income
(in thousands of dollars)

c. Yes, it appears that there is a negative linear correlation. As family income increases, the percent of income donated to charity decreases.

2ab.

Subscriptions
Age (in years)

c. No, it appears that there is no linear correlation between age and subscriptions.

3ab.

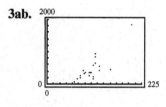

c. Yes, there appears to be a positive linear relationship between budget and worldwide gross.

4a. $n = 6$

b.

x	y	xy	x^2	y^2
42	9	378	1764	81
48	10	480	2304	100
50	8	400	2500	64
59	5	295	3481	25
65	6	390	4225	36
72	3	216	5184	9
$\Sigma x = 336$	$\Sigma y = 41$	$\Sigma xy = 2159$	$\Sigma x^2 = 19{,}458$	$\Sigma y^2 = 315$

c.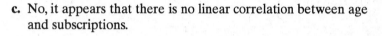

$$r = \frac{n\Sigma xy - (\Sigma x)(\Sigma y)}{\sqrt{n\Sigma x^2 - (\Sigma x)^2}\sqrt{n\Sigma y^2 - (\Sigma y)^2}} = \frac{6(2159) - (336)(41)}{\sqrt{6(19{,}458) - (336)^2}\sqrt{6(315) - (41)^2}}$$

$$= \frac{-822}{\sqrt{3852}\sqrt{209}} \approx -0.916$$

d. Because r is close to -1, there appears to be a strong negative linear correlation between income level and donating percent.

5a. Enter the data.

b. $r \approx 0.838$

c. Because r is close to 1, there appears to be a strong positive linear correlation between budgets and worldwide grosses.

6a. $n = 6$

b. $\alpha = 0.01$

c. 0.917

d. Because $|r| \approx 0.916 < 0.917$, the correlation is not significant.

e. There is not enough evidence to conclude that there is a significant linear correlation between income level and the donating percent.

7a. $H_0: \rho = 0; H_a: \rho \neq 0$

b. $\alpha = 0.01$

c. d.f. $= n - 2 = 23$

d. ± 2.807; Reject H_0 if $t < -2.807$ or $t > 2.807$.

e. $t = \dfrac{r}{\sqrt{\dfrac{1 - r^2}{n - 2}}} = \dfrac{0.83773}{\sqrt{\dfrac{1 - (0.83773)^2}{25 - 2}}} = \dfrac{0.83773}{\sqrt{0.012966}} \approx 7.357$

f. Reject H_0.

g. There is enough evidence in the sample to conclude that a significant linear correlation exists.

9.1 EXERCISE SOLUTIONS

1. $r = -0.925$ represents a stronger correlation because it is closer to -1 than $r = 0.834$ is to $+1$.

3. A table can be used to compare r with a critical value or a hypotheses test can be performed using a t-test.

5. Negative linear correlation   **7.** No linear correlation

9. (c), You would expect a positive linear correlation between age and income.

11. (b), You would expect a negative linear correlation between age and balance on student loans.

13. Explanatory variable: Amount of water consumed

Response variable: Weight loss

15. (a)

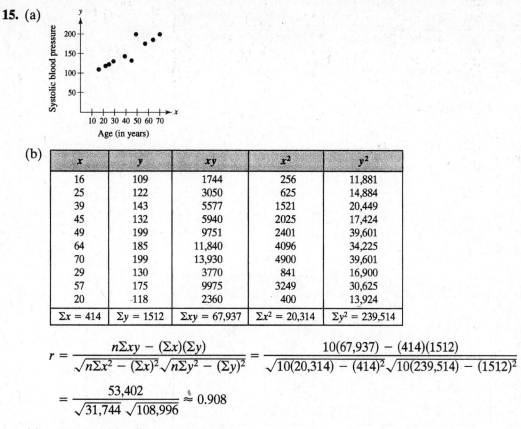

(b)

x	y	xy	x^2	y^2
16	109	1744	256	11,881
25	122	3050	625	14,884
39	143	5577	1521	20,449
45	132	5940	2025	17,424
49	199	9751	2401	39,601
64	185	11,840	4096	34,225
70	199	13,930	4900	39,601
29	130	3770	841	16,900
57	175	9975	3249	30,625
20	118	2360	400	13,924
$\Sigma x = 414$	$\Sigma y = 1512$	$\Sigma xy = 67{,}937$	$\Sigma x^2 = 20{,}314$	$\Sigma y^2 = 239{,}514$

$$r = \frac{n\Sigma xy - (\Sigma x)(\Sigma y)}{\sqrt{n\Sigma x^2 - (\Sigma x)^2}\sqrt{n\Sigma y^2 - (\Sigma y)^2}} = \frac{10(67{,}937) - (414)(1512)}{\sqrt{10(20{,}314) - (414)^2}\sqrt{10(239{,}514) - (1512)^2}}$$

$$= \frac{53{,}402}{\sqrt{31{,}744}\,\sqrt{108{,}996}} \approx 0.908$$

(c) Strong positive linear correlation

17. (a)

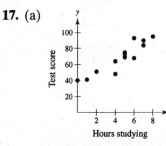

(b)

x	y	xy	x^2	y^2
0	40	0	0	1600
1	41	41	1	1681
2	51	102	4	2601
4	48	192	16	2304
4	64	256	16	4096
5	69	345	25	4761
5	73	365	25	5329
5	75	375	25	5625
6	68	408	36	4624
6	93	558	36	8649
7	84	588	49	7056
7	90	630	49	8100
8	95	760	64	9025
$\Sigma x = 60$	$\Sigma y = 891$	$\Sigma xy = 4620$	$\Sigma x^2 = 346$	$\Sigma y^2 = 65{,}451$

$$r = \frac{n\Sigma xy - (\Sigma x)(\Sigma y)}{\sqrt{n\Sigma x^2 - (\Sigma x)^2}\sqrt{n\Sigma y^2 - (\Sigma y)^2}} = \frac{13(4620) - (60)(891)}{\sqrt{13(346) - (60)^2}\sqrt{13(65,451) - (891)^2}}$$

$$= \frac{6600}{\sqrt{898}\sqrt{56,982}} \approx 0.923$$

(c) Strong positive linear correlation

19. (a)

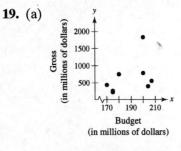

(b)

x	y	xy	x^2	y^2
207	553	114,471	42,849	305,809
204	391	79,764	41,616	152,881
200	1835	367,000	40,000	3.37×10^6
200	784	156,800	40,000	614,656
180	749	134,820	32,400	561,001
175	218	38,150	30,625	47,524
175	255	44,625	30,625	65,025
170	433	73,610	28,900	187,489
$\Sigma x = 1511$	$\Sigma y = 5218$	$\Sigma xy = 1,009,240$	$\Sigma x^2 = 287,015$	$\Sigma y^2 = 5,301,610$

$$r = \frac{n\Sigma xy - (\Sigma x)(\Sigma y)}{\sqrt{n\Sigma x^2 - (\Sigma x)^2}\sqrt{n\Sigma y^2 - (\Sigma y)^2}}$$

$$= \frac{8(1,009,240) - (1511)(5218)}{\sqrt{8(287,015) - (1511)^2}\sqrt{8(5,301,610) - (5218)^2}}$$

$$= \frac{189,522}{\sqrt{12,999}\sqrt{15,185,356}} \approx 0.427$$

(c) Weak positive linear correlation

21. (a)

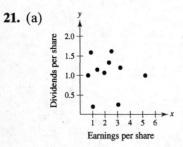

(b)

x	y	xy	x²	y²
2.34	1.33	3.112	5.476	1.769
1.96	1.07	2.097	3.842	1.145
1.39	1.15	1.599	1.932	1.323
3.07	0.25	0.768	9.425	0.063
0.65	1.00	0.650	0.423	1.000
5.21	1.00	5.210	27.144	1.000
0.88	1.59	1.399	0.774	2.528
3.23	1.20	3.876	10.433	1.440
2.54	1.62	4.115	6.452	2.624
1.03	0.20	0.206	1.061	0.040
$\Sigma x = 22.3$	$\Sigma y = 10.41$	$\Sigma xy = 23.031$	$\Sigma x^2 = 66.961$	$\Sigma y^2 = 12.931$

$$r = \frac{n\Sigma xy - (\Sigma x)(\Sigma y)}{\sqrt{n\Sigma x^2 - (\Sigma x)^2}\sqrt{n\Sigma y^2 - (\Sigma y)^2}} = \frac{10(23.031) - (22.3)(10.41)}{\sqrt{10(66.961) - (22.3)^2}\sqrt{10(12.931) - (10.41)^2}}$$

$$= \frac{-1.833}{\sqrt{172.32}\sqrt{20.942}} = -0.031$$

(c) No linear correlation

23. $r \approx 0.623$

$n = 8$ and $\alpha = 0.01$

$cv = 0.834$

$|r| \approx 0.623 < 0.834 \Rightarrow$ The correlation is not significant.

or

$H_0: \rho = 0$; and $H_a: \rho \neq 0$

$\alpha = 0.01$

d.f. $= n - 2 = 6$

$cu = \pm 3.703$; Reject H_0 if $t < -3.707$ or $t > 3.707$.

$$t = \frac{r}{\sqrt{\dfrac{1 - r^2}{n - 2}}} = \frac{0.623}{\sqrt{\dfrac{1 - (0.623)^2}{8 - 2}}} = \frac{0.623}{\sqrt{0.10198}} = 1.951$$

Fail to reject H_0. There is not enough evidence to conclude that there is a significant linear correlation between vehicle weight and the variability in braking distance.

25. $r \approx 0.923$

$n = 8$ and $\alpha = 0.01$

$cv = 0.834$

$|r| \approx 0.923 > 0.834 \Rightarrow$ The correlation is significant.

or

$H_0: \rho = 0$; $H_a: \rho \neq 0$

$\alpha = 0.01$

d.f. $= n - 2 = 11$

$cu = \pm 3.106$; Reject H_0 if $t < -3.106$ or $t > 3.106$.

$r \approx 0.923$

$$t = \frac{r}{\sqrt{\dfrac{1 - r^2}{n - 2}}} = \frac{0.923}{\sqrt{\dfrac{1 - (0.923)^2}{13 - 2}}} = \frac{0.923}{\sqrt{0.01346}} \approx 7.955$$

Reject H_0. There is enough evidence to conclude that a significant linear correlation exists.

27. $r \approx -0.030$

$n = 10$ and $\alpha = 0.01$

$cv = 0.765$

$|r| \approx 0.030 < 0.765 \Rightarrow$ The correlation is not significant.

or

$H_0: \rho = 0; H_a: \rho \neq 0$

$\alpha = 0.01$

d.f. $= n - 2 = 8$

$cu = \pm 3.355$; Reject H_0 if $t < -3.355$ or $t > 3.355$.

$r \approx -0.030$

$$t = \frac{r}{\sqrt{\dfrac{1 - r^2}{n - 2}}} = \frac{-0.030}{\sqrt{\dfrac{1 - (-0.030)^2}{10 - 2}}}$$

$$= \frac{-0.030}{\sqrt{0.125}} = -0.085$$

Fail to reject H_0. There is not enough evidence at the 1% significance level to conclude there is a significant linear correlation between earnings per share and dividends per share.

29. The correlation coefficient remains unchanged when the x-values and y-values are switched.

31. Answers will vary.

9.2 LINEAR REGRESSION

9.2 Try It Yourself Solutions

1a. $n = 6$

x	y	xy	x^2
42	9	378	1764
48	10	480	2304
50	8	400	2500
59	5	295	3481
65	6	390	4225
72	3	216	5184
$\Sigma x = 336$	$\Sigma y = 41$	$\Sigma xy = 2159$	$\Sigma x^2 = 19{,}458$

b. $m = \dfrac{n\Sigma xy - (\Sigma x)(\Sigma y)}{n\Sigma x^2 - (\Sigma x)^2} = \dfrac{6(2159) - (336)(41)}{6(19{,}458) - (336)^2} = \dfrac{-822}{3852} \approx -0.2133956$

c. $b = \bar{y} - m\bar{x} = \left(\dfrac{41}{6}\right) - (-0.2133956)\left(\dfrac{336}{6}\right) \approx 18.7837$

d. $\hat{y} = -0.213x + 18.783$

2a. Enter the data.

b. $m \approx 10.93477; b \approx -710.61551$

c. $\hat{y} = 10.935x - 710.616$

3a. (1) $\hat{y} = 12.481(2) + 33.683$ (2) $\hat{y} = 12.481(3.32) + 33.683$

  b. (1) 58.645   (2) 75.120

  c. (1) 58.645 minutes   (2) 75.120 minutes

9.2 EXERCISE SOLUTIONS

1. c   **3.** d   **5.** g   **7.** h   **9.** c   **11.** a

13.

x	y	xy	x²
764	55	42,020	583,696
625	47	29,375	390,625
520	51	26,520	270,400
510	28	14,280	260,100
492	39	19,188	242,064
484	34	16,456	234,256
450	33	14,850	202,500
430	31	13,330	184,900
410	40	16,400	168,100
$\Sigma x = 4685$	$\Sigma y = 358$	$\Sigma xy = 192{,}419$	$\Sigma x^2 = 2{,}536{,}641$

$$m = \frac{n\Sigma xy - (\Sigma x)(\Sigma y)}{n\Sigma x^2 - (\Sigma x)^2} = \frac{9(192{,}419) - (4685)(358)}{9(2{,}536{,}641) - (4685)^2} = \frac{54{,}541}{880{,}544} \approx 0.06194 \approx 0.062$$

$$b = \bar{y} - m\bar{x} = \left(\frac{358}{9}\right) - (0.06194)\left(\frac{4685}{9}\right) \approx 7.535$$

$$\hat{y} = 0.062x + 7.535$$

(a) $\hat{y} = 0.062(500) + 7.535 \approx 38.535 \approx 39$ stories

(b) $\hat{y} = 0.062(650) + 7.535 \approx 47.835 \approx 48$ stories

(c) It is not meaningful to predict the value of y for $x = 310$ because $x = 310$ is outside the range of the original data.

(d) $\hat{y} = 0.062(725) + 7.535 \approx 52.485 \approx 52$ stories

15.

x	y	xy	x²
0	40	0	0
1	41	41	1
2	51	102	4
4	48	192	16
4	64	256	16
5	69	345	25
5	73	365	25
5	75	375	25
6	68	408	36
6	93	558	36
7	84	588	49
7	90	630	49
8	95	760	64
$\Sigma x = 60$	$\Sigma y = 891$	$\Sigma xy = 4620$	$\Sigma x^2 = 346$

$$m = \frac{n\Sigma xy - (\Sigma x)(\Sigma y)}{n\Sigma x^2 - (\Sigma x)^2} = \frac{13(4620) - (60)(891)}{13(346) - (60)^2} = \frac{6600}{898} \approx 7.350$$

$$b = \bar{y} - m\bar{x} = \left(\frac{60}{13}\right) - (7.350)\left(\frac{891}{13}\right) \approx 34.617$$

$\hat{y} = 7.350x + 34.617$

(a) $\hat{y} = 7.350(3) + 34.617 \approx 56.7$

(b) $\hat{y} = 7.350(6.5) + 34.617 \approx 82.4$

(c) It is not meaningful to predict the value of y for $x = 13$ because $x = 13$ is outside the range of the original data.

(d) $\hat{y} = 7.350(4.5) + 34.617 \approx 67.7$

17.

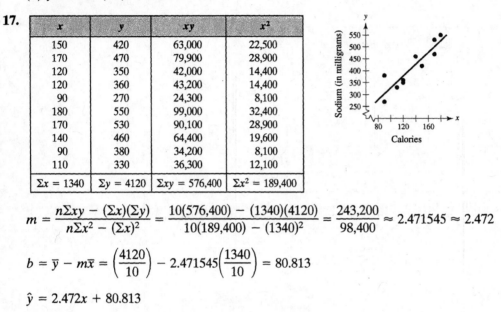

x	y	xy	x^2
150	420	63,000	22,500
170	470	79,900	28,900
120	350	42,000	14,400
120	360	43,200	14,400
90	270	24,300	8,100
180	550	99,000	32,400
170	530	90,100	28,900
140	460	64,400	19,600
90	380	34,200	8,100
110	330	36,300	12,100
$\Sigma x = 1340$	$\Sigma y = 4120$	$\Sigma xy = 576,400$	$\Sigma x^2 = 189,400$

$$m = \frac{n\Sigma xy - (\Sigma x)(\Sigma y)}{n\Sigma x^2 - (\Sigma x)^2} = \frac{10(576,400) - (1340)(4120)}{10(189,400) - (1340)^2} = \frac{243,200}{98,400} \approx 2.471545 \approx 2.472$$

$$b = \bar{y} - m\bar{x} = \left(\frac{4120}{10}\right) - 2.471545\left(\frac{1340}{10}\right) = 80.813$$

$$\hat{y} = 2.472x + 80.813$$

(a) $\hat{y} = 2.472(170) + 80.813 = 501.053$ milligrams

(b) $\hat{y} = 2.472(100) + 80.813 = 328.213$ milligrams

(c) $\hat{y} = 2.472(140) + 80.813 = 426.893$ milligrams

(d) It is not meaningful to predict the value of y for $x = 210$ because $x = 210$ is outside the range of the original data.

19.

x	y	xy	x^2
8.5	66.0	561.00	72.25
9.0	68.5	616.50	81.00
9.0	67.5	607.50	81.00
9.5	70.0	665.00	90.25
10.0	70.0	700.00	100.00
10.0	72.0	720.00	100.00
10.5	71.5	750.75	110.25
10.5	69.5	729.75	110.25
11.0	71.5	786.50	121.00
11.0	72.0	792.00	121.00
11.0	73.0	803.00	121.00
12.0	73.5	882.00	144.00
12.0	74.0	888.00	144.00
12.5	74.0	925.00	156.25
$\Sigma x = 146.5$	$\Sigma y = 993.0$	$\Sigma xy = 10,427.0$	$\Sigma x^2 = 1552.3$

$$m = \frac{n\Sigma xy - (\Sigma x)(\Sigma y)}{n\Sigma x^2 - (\Sigma x)^2} = \frac{14(10,427.0) - (146.5)(993.0)}{14(1552.3) - (146.5)^2} = \frac{503.5}{269.95} \approx 1.870$$

$$b = \bar{y} - m\bar{x} = \left(\frac{993.0}{14}\right) - (1.870)\left(\frac{146.5}{14}\right) \approx 51.360$$

$$\hat{y} = 1.870x + 51.360$$

(a) $\hat{y} = 1.870(11.5) + 51.360 \approx 72.865$ inches

(b) $\hat{y} = 1.870(8.0) + 51.360 \approx 66.32$ inches

(c) It is not meaningful to predict the value of y for $x = 15.5$ because $x = 15.5$ is outside the range of the original data.

(d) $\hat{y} = 1.870(10.0) + 51.360 \approx 70.06$ inches

21.

x	y	xy	x^2
5720	2.19	12,527	32,718,400
4050	1.36	5508	16,402,500
6130	2.58	15,815	37,576,900
5000	1.74	8700	25,000,000
5010	1.78	8917.8	25,100,100
4270	1.69	7216.3	18,232,900
5500	1.80	9900	30,250,000
5550	1.87	10,379	30,802,500
$\Sigma x = 41,230$	$\Sigma y = 15.01$	$\Sigma xy = 78,962.8$	$\Sigma x^2 = 216,083,300$

$$m = \frac{n\Sigma xy - (\Sigma x)(\Sigma y)}{n\Sigma x^2 - (\Sigma x)^2} = \frac{8(78,962.8) - (41,230)(15.01)}{8(216,083,300) - (41,230)^2} = \frac{12,833.7}{28,753,500} = 0.000447$$

$$b = \bar{y} - m\bar{x} = \left(\frac{15.01}{8}\right) - (0.000447)\left(\frac{41,230}{8}\right) = -0.425$$

$$\hat{y} = 0.000447x - 0.425$$

(a) $\hat{y} = 0.000447(4500) - 0.425 = 1.587$ feet

(b) $\hat{y} = 0.000447(6000) - 0.425 = 2.257$ feet

(c) It is not meaningful to predict the value of y for $x = 7500$ because $x = 7500$ is outside the range of the original data.

(d) $\hat{y} = 0.000447(5750) - 0.425 = 2.145$ feet

23. Substitute a value x into the equation of a regression line and solve for y.

25. Strong positive linear correlation; As the ages of the engineers increase, the salaries of the engineers tend to increase.

27. It is not meaningful to predict y for $x = 74$ because $x = 74$ is outside the range of the original data.

29. In general, as age increases; salary increases until age 61 when salary decreases. (Answers will vary.)

31. (a) 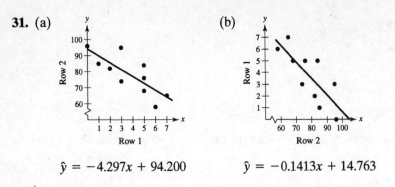 (b)

$\hat{y} = -4.297x + 94.200$ $\hat{y} = -0.1413x + 14.763$

(c) The slope of the line keeps the same sign, but the values of m and b change.

33. (a) $m = 0.139$
    $b = 21.024$
    $\hat{y} = 0.139x + 21.024$

(b)

(c)

x	y	$\hat{y} = 0.139x + 21.024$	$y - \hat{y}$
38	24	26.306	−2.306
34	22	25.750	−3.750
40	27	26.584	0.416
46	32	27.418	4.582
43	30	27.001	2.999
48	31	27.696	3.304
60	27	29.364	−2.364
55	26	28.669	−2.669
52	28	28.252	−0.252

(d) The regression line may not be a good model for the data because the data do not fluctuate about 0.

35. (a)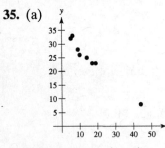

(b) The point (44, 8) may be an outlier.

(c) Excluding the point (44, 8) $\Rightarrow \hat{y} = -0.711x + 35.263$. The point (44, 8) is not influential because using all 8 points $\Rightarrow \hat{y} = -0.607x + 34.160$.

37. $m = 654.536$

$b = -1214.857$

$\hat{y} = 654.536x - 1214.857$

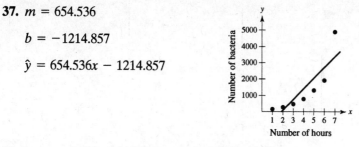

39. $\log y = 0.233x + 1.969 \Rightarrow y = 10^{0.233x + 1.969}$

$\Rightarrow y = 93.111(10)^{0.233x}$

41. $m = -78.929$

$b = 576.179$

$\hat{y} = -78.929x + 576.179$

43. $\log y = m(\log x) + b \Rightarrow y = 10^{m \log x + b} = 10^b 10^{m \log x}$

$= 10^{2.893} 10^{-1.251 \log x} = 781.628 \cdot x^{-1.251}$

45. $y = a + b \ln x = 25.035 + 19.599 \ln x$

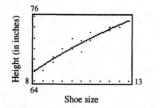

47. The logarithmic equation is a better model for the data. The logarithmic equation's error is smaller.

9.3 MEASURES OF REGRESSION AND PREDICTION INTERVALS

9.3 Try It Yourself Solutions

1a. $r = 0.979$

b. $r^2 = (0.979)^2 = 0.958$

c. 95.8% of the variation in the times is explained.
4.2% of the variation is unexplained.

2a.

x_i	y_i	$\hat{y}_i$	$(y_i - \hat{y}_i)^2$
15	26	28.386	5.693
20	32	35.411	11.635
20	38	35.411	6.703
30	56	49.461	42.759
40	54	63.511	90.459
45	78	70.536	55.711
50	80	77.561	5.949
60	88	91.611	13.039
			$\Sigma = 231.948$

b. $n = 8$

c. $s_e = \sqrt{\dfrac{\Sigma(y_i - \hat{y}_i)^2}{n - 2}} = \sqrt{\dfrac{231.948}{6}} \approx 6.218$

d. The standard error of estimate of the weekly sales for a specific radio ad time is about \$621.80.

3a. $n = 8$, d.f. = 6, $t_c = 2.447$, $s_e \approx 10.290$

b. $\hat{y} = 50.729x + 104.061 = 50.729(2.5) + 104.061 \approx 230.884$

c. $E = t_c s_e \sqrt{1 + \dfrac{1}{n} + \dfrac{n(x - \bar{x})^2}{n(\Sigma x^2) - (\Sigma x)^2}} = (2.447)(10.290)\sqrt{1 + \dfrac{1}{8} + \dfrac{8(2.5 - 1.975)^2}{8(32.44) - (15.8)^2}}$

$= (2.447)(10.290)\sqrt{1.34818} \approx 29.236$

d. $\hat{y} \pm E \rightarrow (201.648, 260.120)$

e. You can be 95% confident that the company sales will be between \$201,648 and \$260,120 when advertising expenditures are \$2500.

9.3 EXERCISE SOLUTIONS

1. Total variation $= \Sigma(y_i - \bar{y})^2$; the sum of the squares of the differences between the y-values of each ordered pair and the mean of the y-values of the ordered pairs.

3. Unexplained variation $= \Sigma(y_i - \hat{y}_i)^2$; the sum of the squares of the differences between the observed y-values and the predicted y-values.

5. $r^2 = (0.350)^2 \approx 0.123$

12.3% of the variation is explained. 87.7% of the variation is unexplained.

7. $r = (-0.891)^2 \approx 0.794$

79.4% of the variation is explained. 20.6% of the variation is unexplained.

9. (a) $r^2 = \dfrac{\Sigma(\hat{y}_i - \bar{y})^2}{\Sigma(y_i - \bar{y})^2} \approx 0.233$

23.3% of the variation in proceeds can be explained by the variation in the number of issues and 76.7% of the variation is unexplained.

(b) $s_e = \sqrt{\dfrac{\Sigma(y_i - \hat{y}_i)^2}{n - 2}} = \sqrt{\dfrac{2,356,140,670}{10}} \approx 15,349.725$

The standard error of estimate of the proceeds for a specific number of issues is about \$15,349,725,000.

11. (a) $r^2 = \dfrac{\Sigma(\hat{y}_i - \bar{y})^2}{\Sigma(y_i - \bar{y})^2} \approx 0.981$

98.1% of the variation in sales can be explained by the variation in the total square footage and 1.9% of the variation is unexplained.

(b) $s_e = \sqrt{\dfrac{\Sigma(y_i - \hat{y}_i)^2}{n - 2}} = \sqrt{\dfrac{8413.958}{9}} \approx 30.576$

The standard error of estimate of the sales for a specific total square footage is about 30,576,000,000.

13. (a) $r^2 = \dfrac{\Sigma(\hat{y}_i - \bar{y})^2}{\Sigma(y_i - \bar{y})^2} \approx 0.994$

99.4% of the variation in the median weekly earnings of female workers can be explained by the variation in the median weekly earnings of male workers and 0.6% of the variation is unexplained.

(b) $s_e = \sqrt{\dfrac{\Sigma(y_i - \hat{y}_i)^2}{n - 2}} = \sqrt{\dfrac{22.178}{3}} \approx 2.719$

The standard error of estimate of the median weekly earnings of female workers for a specific median weekly earnings of male workers is about $2.72.

15. (a) $r^2 = \dfrac{\Sigma(\hat{y}_i - \bar{y})^2}{\Sigma(y_i - \bar{y})^2} \approx 0.994$

99.4% of the variation in the money spent can be explained by the variation in the money raised and 0.6% of the variation is unexplained.

(b) $s_e = \sqrt{\dfrac{\Sigma(y_i - \hat{y}_i)^2}{n - 2}} = \sqrt{\dfrac{2136.181}{6}} \approx 18.869$

The standard error of estimate of the money spent for a specified amount of money raised is about $18,869,000.

17. $n = 12$, d.f. $= 10$, $t_c = 2.228$, $s_e = 15,349.725$

$\hat{y} = 40.049x + 21,843.09 = 40.049(712) + 21,843.09 \approx 50,357.978$

$E = t_c s_e \sqrt{1 + \dfrac{1}{n} + \dfrac{n(x - \bar{x})^2}{n(\Sigma x^2) - (\Sigma x)^2}} = (2.228)(15,349.725)\sqrt{1 + \dfrac{1}{12} + \dfrac{12(712 - 3806/12)^2}{12(1,652,648) - (3806)^2}}$

$\quad = (2.228)(15,349.725)\sqrt{1.43325} \approx 40,942.727$

$\hat{y} \pm E \rightarrow (9415.251, 91,300.705) \rightarrow (\$9,415,251,000, \$91,300,705,000)$

You can be 95% confident that the proceeds will be between $9,415,251,000 and $91,300,705,000 when the number of initial offerings is 712.

19. $n = 11$, d.f. $= 9$, $t_c = 1.833$, $s_e \approx 30.576$

$\hat{y} = 548.448x - 1881.694 = 548.448(4.5) - 1881.694 \approx 590.822$

$E = t_c s_e \sqrt{1 + \dfrac{1}{n} + \dfrac{n(x - \bar{x})^2}{n(\Sigma x^2) - (\Sigma x)^2}} = (1.833)(30.576)\sqrt{1 + \dfrac{1}{11} + \dfrac{11[4.5 - (61.2/11)]^2}{11(341.9) - (61.2)^2}}$

$\quad = (1.833)(30.576)\sqrt{1.89586} \approx 77.170$

$\hat{y} \pm E \rightarrow (513.652, \ 677.992) \rightarrow (\$513,652,000,000, \$667,992,000,000)$

You can be 90% confident that the sales will be between $513,652,000,000 and $667,992,000,000 when the total square footage is 4.5 billion.

21. $n = 5$, d.f. = 3, $t_c = 5.841$, $s_e \approx 2.719$

$\hat{y} = 1.369x - 402.687 = 1.369(650) - 402.687 \approx 487.163$

$$E = t_c s_e \sqrt{1 + \frac{1}{n} + \frac{n(x - \bar{x})^2}{n(\Sigma x^2) - (\Sigma x)^2}} = (5.841)(2.719) \sqrt{1 + \frac{1}{5} + \frac{5[650 - (3479/5)]^2}{5(2,422,619) - (3479)^2}}$$

$$= (5.841)(2.719) \sqrt{2.28641} \approx 24.014$$

$\hat{y} \pm E \rightarrow (463.149, 511.177)$

You can be 99% confident that the median earnings of female workers will be between $463.15 and $511.18 when the median weekly earnings of male workers is $650.

23. $n = 8$, d.f. = 6, $t_c = 2.447$, $s_e \approx 18.869$

$\hat{y} = 0.943x - 21.541 = 0.943(775.8) + 21.541 \approx 753.120$

$$E = t_c s_e \sqrt{1 + \frac{1}{n} + \frac{n(x - \bar{x})^2}{n(\Sigma x^2) - (\Sigma x)^2}} = (2.447)(18.869) \sqrt{1 + \frac{1}{8} + \frac{8[775.8 - (6666.2/8)]^2}{8(5,932,282.32) - (6666.2)^2}}$$

$$= (2.447)(18.869) \sqrt{1.13375} \approx 49.163$$

$\hat{y} \pm E \rightarrow (\$703.957 \text{ million}, \$802.283 \text{ million})$

You can be 95% confident that the money spent in congressional campaigns will be between $703.957 million and $802.283 million when the money raised is $775.8 million.

25.

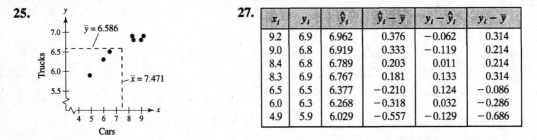

27.

x_i	y_i	$\hat{y}_i$	$\hat{y}_i - \bar{y}$	$y_i - \hat{y}_i$	$y_i - \bar{y}$
9.2	6.9	6.962	0.376	−0.062	0.314
9.0	6.8	6.919	0.333	−0.119	0.214
8.4	6.8	6.789	0.203	0.011	0.214
8.3	6.9	6.767	0.181	0.133	0.314
6.5	6.5	6.377	−0.210	0.124	−0.086
6.0	6.3	6.268	−0.318	0.032	−0.286
4.9	5.9	6.029	−0.557	−0.129	−0.686

29. $r^2 \approx 0.919$; About 91.9% of the variation in the median age of trucks can be explained by the variation in the median age of cars, and 8.1% of the variation is unexplained.

31. $\hat{y} = 0.217x + 4.966 = 0.217(8.6) + 4.966 \approx 6.832$

$$E = t_c s_e \sqrt{1 + \frac{1}{n} + \frac{n(x - \bar{x})^2}{n(\Sigma x^2) - (\Sigma x)^2}} = (2.571)(0.117) \sqrt{1 + \frac{1}{7} + \frac{7[8.6 - (52.3/7)]^2}{7(407.35) - (52.3)^2}}$$

$$= (2.571)(0.117) \sqrt{1.21961} \approx 0.332$$

$\hat{y} \pm E \rightarrow (6.500, 7.164)$

33. $cv = \pm 3.707$

$m = -0.205$

$se = 0.554$

$$t = \frac{m}{s_e} \sqrt{\Sigma x^2 - \frac{(\Sigma x)^2}{n}} = \frac{-0.205}{0.554} \sqrt{838.55 - \frac{(79.5)^2}{8}} = -2.578$$

Fail to reject H_0: $M = 0$

35. $E = t_c s_e \sqrt{\dfrac{1}{n} + \dfrac{\bar{x}^2}{\Sigma x^2 - [(\Sigma x)^2/n]}} = 2.447(10.290)\sqrt{\dfrac{1}{8} + \dfrac{\left(\frac{15.8}{8}\right)^2}{(32.44) - \frac{(15.8)^2}{8}}} = 45.626$

$b \pm E \Rightarrow 104.061 \pm 45.626 = (58.435, 149.687)$

$E = \dfrac{t_c s_e}{\sqrt{\Sigma x^2 - \dfrac{(\Sigma x)^2}{n}}} = \dfrac{2.447(10.290)}{\sqrt{32.44 - \dfrac{(15.8)^2}{8}}} = 22.658$

$m \pm E \Rightarrow 50.729 \pm 22.658 \Rightarrow (28.071, 73.387)$

9.4 MULTIPLE REGRESSION

9.4 Try It Yourself Solutions

1a. Enter the data.   **b.** $\hat{y} = 46.385 + 0.540x_1 - 4.897x_2$

2ab. (1) $\hat{y} = 46.385 + 0.540(89) - 4.897(1)$
   (2) $\hat{y} = 46.385 + 0.540(78) - 4.897(3)$
   (3) $\hat{y} = 46.385 + 0.540(83) - 4.897(2)$

c. (1) $\hat{y} = 89.548$ (2) $\hat{y} = 73.814$ (3) $\hat{y} = 81.411$

d. (1) 90   (2) 74   (3) 81

9.4 EXERCISE SOLUTIONS

1. $\hat{y} = 640 - 0.105x_1 + 0.124x_2$

   (a) $\hat{y} = 640 - 0.105(13,500) + 0.124(12,000) = 710.5$ pounds

   (b) $\hat{y} = 640 - 0.105(15,000) + 0.124(13,500) = 739$ pounds

   (c) $\hat{y} = 640 - 0.105(14,000) + 0.124(13,500) = 844$ pounds

   (d) $\hat{y} = 640 - 0.105(14,500) + 0.124(13,000) = 729.5$ pounds

3. $\hat{y} = -52.2 + 0.3x_1 + 4.5x_2$

   (a) $\hat{y} = -52.2 + 0.3(70) + 4.5(8.6) = 7.5$ cubic feet

   (b) $\hat{y} = -52.2 + 0.3(65) + 4.5(11.0) = 16.8$ cubic feet

   (c) $\hat{y} = -52.2 + 0.3(83) + 4.5(17.6) = 51.9$ cubic feet

   (d) $\hat{y} = -52.2 + 0.3(87) + 4.5(19.6) = 62.1$ cubic feet

5. $\hat{y} = -2518.4 + 126.8x_1 + 66.4x_2$

   (a) $s = 28.489$

   (b) $r^2 = 0.985$

   (c) The standard error of estimate of the predicted sales given a specific total square footage and number of shopping centers is \$28.489 billion. The multiple regression model explains 98.5% of the variation in y.

7. $n = 11, k = 2, r^2 = 0.985$

$$r_{adj}^2 = 1 - \left[\frac{(1 - r^2)(n - 1)}{n - k - 1}\right] = 0.981$$

98.1% of the variation in y can be explained by the relationships between variables.

$r_{adj}^2 < r^2$

CHAPTER 9 REVIEW EXERCISE SOLUTIONS

1.

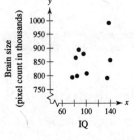

Age (in years)

$r \approx -0.939$; strong negative linear correlation; milk production decreases as age increases

3.

$r \approx 0.361$; weak positive linear correlation; brain size increases as IQ increases.

5. $H_0: \rho = 0; H_a: \rho \neq 0$

$\alpha = 0.01$, d.f. $= n - 2 = 24$

$t_0 = 2.797$

$$t = \frac{r}{\sqrt{\dfrac{1 - r^2}{n - 2}}} = \frac{0.24}{\sqrt{\dfrac{1 - (0.24)^2}{26 - 2}}} = \frac{0.24}{\sqrt{0.03927}} = 1.211$$

Fail to reject H_0. There is not enough evidence to conclude that a significant linear correlation exists.

7. $H_0: \rho = 0; H_a: \rho \neq 0$

$\alpha = 0.05$, d.f. $= n - 2 = 6$

$t_0 = \pm 2.447$

$$t = \frac{r}{\sqrt{\dfrac{1 - r^2}{n - 2}}} = \frac{-0.939}{\sqrt{\dfrac{1 - (-0.939)^2}{8 - 2}}} = \frac{-0.939}{\sqrt{0.01971}} = -6.688$$

Reject H_0. There is enough evidence to conclude that there is a significant linear correlation between the age of a cow and its milk production.

9. $H_0: \rho = 0; H_a: \rho \neq 0$

$\alpha = 0.01$, d.f. $= n - 2 = 7$

$t_0 = \pm 3.499$

$$t = \frac{r}{\sqrt{\dfrac{1 - r^2}{n - 2}}} = \frac{0.361}{\sqrt{\dfrac{1 - (0.361)^2}{7}}} = 1.024$$

Fail to reject H_0. There is not enough evidence to conclude a significant linear correlation exists between brain size and IQ.

11. $\hat{y} = 0.757x + 21.525$

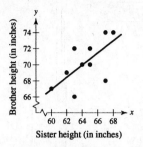

13. $\hat{y} = -0.086x + 10.450$

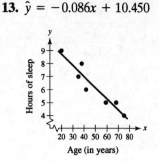

$r \approx 0.688$ (Moderate positive linear correlation)

15. (a) $\hat{y} = 0.757(61) + 21.525 = 67.702 \approx 68$ inches

(b) $\hat{y} = 0.757(66) + 21.525 = 71.487 \approx 71$ inches

(c) Not meaningful because $x = 75$ inches is outside range of data.

(d) Not meaningful because $x = 50$ inches is outside range of data.

17. (a) 8.902 hours

(b) $\hat{y} = 10.450 = 0.086x = 10.450 - 0.086(25) = 8.3$ hours

(c) Not meaningful because $x = 85$ years is outside range of data.

(d) $\hat{y} = 10.450 - 0.086x = 10.450 - 0.086(50) = 6.15$ hours

19. $r^2 = (-0.553)^2 = 0.306$

30.6% of the variation in y is explained.
69.4% of the variation in y is unexplained.

21. $r^2 = (0.181)^2 = 0.033$

3.3% of the variation in y is explained.
96.7% of the variation in y is unexplained.

23. (a) $r^2 = 0.897$

89.7% of the variation in y is explained.
10.3% of the variation in y is unexplained.

(b) $s_e = 568.011$

The standard error of estimate of the cooling capacity for a specific living area is 568.0 Btu per hour.

25. $\hat{y} = 0.757(64) + 21.525 = 69.973$

$$E = t_c s_e \sqrt{1 + \frac{1}{n} + \frac{n(x - \bar{x})^2}{n\Sigma x^2 - (\Sigma x)^2}} \approx (1.833)(2.206)\sqrt{1 + \frac{1}{11} + \frac{11(64 - 64.727)^2}{11(46,154) - 712^2}}$$

$$= (1.833)(2.202)\sqrt{1.09866} \approx 4.231$$

$$\hat{y} - E < y < \hat{y} + E$$
$$69.973 - 4.231 < y < 69.973 + 4.231$$
$$65.742 < y < 74.204$$

You can be 90% confident that the height of a male will be between 65.742 inches and 74.204 inches when his sister is 64 inches tall.

27. $\hat{y} = -0.086(45) + 10.450 = 6.580$

$$E = t_c s_e \sqrt{1 + \frac{1}{n} + \frac{n(x - \bar{x})^2}{n\Sigma x^2 - (\Sigma x)^2}} \approx (2.571)(0.623)\sqrt{1 + \frac{1}{7} + \frac{7[45 - (337/7)]^2}{7(18,563) - (337)^2}}$$

$$= (2.571)(0.623)\sqrt{1.14708} = 1.715$$

$$\hat{y} - E < y < \hat{y} + E$$

$$6.580 - 1.715 < y < 6.58 + 1.715$$

$$4.865 < y < 8.295$$

You can be 95% confident that the hours slept will be between 4.865 and 8.295 hours for a person who is 45 years old.

29. $\hat{y} = 3002.991 + 9.468(720) = 9819.95$

$$E = t_c s_e \sqrt{1 + \frac{1}{n} + \frac{n(x - \bar{x})^2}{n\Sigma x^2 - (\Sigma x)^2}} = (3.707)(568)\sqrt{1 + \frac{1}{8} + \frac{8(720 - 499.4)^2}{8(2,182,275) - 3995^2}}$$

$$= (3.707)(568)\sqrt{1.38486} \approx 2477.94$$

$$\hat{y} - E < y < \hat{y} + E$$

$$9819.95 - 2477.94 < y < 9819.95 + 2477.94$$

$$7342.01 < y < 12,297.89$$

You can be 99% confident that the cooling capacity will be between 7342.01 Btu per hour and 12,297.89 Btu per hour when the living area is 720 square feet.

31. $\hat{y} = 3.674 + 1.287x_1 + -7.531x_2$

33. (a) 21.705   (b) 25.21   (c) 30.1   (d) 25.86

CHAPTER 9 QUIZ SOLUTIONS

1.

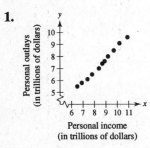

The data appear to have a positive linear correlation. The outlays increase as the incomes increase.

2. $r \approx 0.996 \rightarrow$ Strong positive linear correlation

3. $H_0: \rho = 0$; $H_a: \rho \neq 0$

$\alpha = 0.05$, d.f. $= n - 2 = 9$

$t_0 = \pm 2.262$

$$t = \frac{r}{\sqrt{\dfrac{1 - r^2}{n - 2}}} = \frac{0.996}{\sqrt{\dfrac{1 - (0.996)^2}{11 - 2}}} = \frac{0.996}{\sqrt{0.00088711}} \approx 33.44$$

Reject H_0. There is enough evidence to conclude that a significant correlation exists.

4. $\Sigma y^2 = 616.090$

$\Sigma y = 81.100$

$\Sigma xy = 715.940$

$b = -1.018$

$m = 0.976$

$\hat{y} = 0.976x - 1.018$

Personal income
(in trillions of dollars)

5. $\hat{y} = 0.976(6.2) - 1.018 = \$5.033$ trillion

6. $r^2 = 0.993$

99.3% of the variation in y is explained.

0.7% of the variation in y is unexplained.

7. $s_e = 0.121$

The standard deviation of personal outlays for a specified personal income is \$0.121 trillion.

8. $\hat{y} = 0.976(7.6) - 1.018 = 6.400$

$$E = t_c s_e \sqrt{1 + \frac{1}{n} + \frac{n(x - \bar{x})^2}{n\Sigma x^2 - (\Sigma x)^2}}$$

$$= 2.262(0.120)\sqrt{1 + \frac{1}{11} + \frac{11[7.6 - (94.6/11)]^2}{11(832.5) - 94.6^2}}$$

$$\approx (2.262)(0.120)\sqrt{1.144} \approx 0.290$$

$$\hat{y} - E < y < \hat{y} + E$$

$$6.400 - 0.290 < y < 6.400 + 0.290$$

$$6.110 < y < 6.690 \rightarrow (\$6.110 \text{ trillion}, \$6.690 \text{ trillion})$$

You can be 95% confident that the personal outlays will be between \$6.110 trillion and \$6.690 trillion when personal income is \$7.6 trillion.

9. (a) 1374.762 pounds

(b) 1556.755 pounds

(c) 1183.262 pounds

(d) 1294.385

x_2 has the greater influence on y.

10.1 GOODNESS OF FIT

10.1 Try It Yourself Solutions

1a.

Music	% of Listeners	Expected Frequency
Classical	4%	12
Country	36%	108
Gospel	11%	33
Oldies	2%	6
Pop	18%	54
Rock	29%	87

2a. The expected frequencies are 64, 80, 32, 56, 60, 48, 40, and 20, all of which are at least 5.

b. Claimed Distribution:

Ages	Distribution
0–9	16%
10–19	20%
20–29	8%
30–39	14%
40–49	15%
50–59	12%
60–69	10%
70+	5%

H_0: Distribution of ages is as shown in table above.

H_a: Distribution of ages differs from the claimed distribution.

c. $\alpha = 0.05$ **d.** d.f. $= n - 1 = 7$

e. $\chi_0^2 = 14.067$; Reject H_0 if $\chi^2 > 14.067$.

f.

Ages	Distribution	Observed	Expected	$\dfrac{(O - E)^2}{E}$
0–9	16%	76	64	2.250
10–19	20%	84	80	0.200
20–29	8%	30	32	0.125
30–39	14%	60	56	0.286
40–49	15%	54	60	0.600
50–59	12%	40	48	1.333
60–69	10%	42	40	0.100
70+	5%	14	20	1.800
				6.694

$\chi^2 \approx 6.694$

g. Fail to reject H_0.

h. There is not enough evidence to suppport the claim that the distribution of ages differs from the claimed distribution.

3a. The expected frequencies are 123, 138, and 39, all of which are at least 5.

b. Claimed Distribution:

Response	Distribution
Retirement	41%
Children's college education	46%
Not sure	13%

H_0: Distribution of responses is as shown in table above. (claim)

H_a: Distribution of responses differs from the claimed distribution.

c. $\alpha = 0.01$ **d.** d.f. $= n - 1 = 2$ **e.** $\chi_0^2 = 9.210$; Reject H_0 if $\chi^2 > 9.210$.

f.

Response	Distribution	Observed	Expected	$\dfrac{(O - E)^2}{E}$
Retirement	41%	129	123	0.293
Children's college education	46%	149	138	0.877
Not sure	13%	22	39	7.410
				8.580

$\chi^2 \approx 8.580$

g. Fail to reject H_0.

h. There is not enough evidence to dispute the claimed distribution of responses.

4a. The expected frequencies are 30 for each color.

b. H_0: The distribution of different-colored candies in bags of peanut M&M's is uniform. (claim)

H_1: The distribution of different-colored candies in bags of peanut M&M's is not uniform.

c. $\alpha = 0.05$ **d.** d.f. $= k - 1 = 6 - 1 = 5$ **e.** $\chi_0^2 = 11.071$; Reject H_0 if $\chi^2 > 11.071$.

f.

Color	Observed	Expected	$\dfrac{(O - E)^2}{E}$
Brown	22	30	2.133
Yellow	27	30	0.300
Red	22	30	2.133
Blue	41	30	4.033
Orange	41	30	4.033
Green	27	30	0.300
			$\chi^2 = \sum \dfrac{(O - E)^2}{E} = 12.933$

g. Reject H_0.

h. There is enough evidence to dispute the claimed distribution.

10.1 EXERCISE SOLUTIONS

1. A multinomial experiment is a probability experiment consisting of a fixed number of trials in which there are more than two possible outcomes for each independent trial.

3. $E_i = np_i = (150)(0.3) = 45$

5. $E_i = np_i = (230)(0.25) = 57.5$

7. (a) Claimed distribution:

Response	Distribution
Reward Program	28%
Low Interest Rate	24%
Cash Back	22%
Special Store Discounts	8%
Other	16%

H_0: Distribution of responses is as shown in the table above.

H_a: Distribution of responses differs from the claimed or expected distribution. (claim)

(b) $\chi_0^2 = 9.488$; Reject H_0 if $\chi^2 > 9.488$.

(c)

Response	Distribution	Observed	Expected	$\frac{(O-E)^2}{E}$
Reward Program	28%	112	119	0.412
Low Interest Rate	24%	98	102	0.157
Cash Back	22%	107	93.5	1.949
Special Store Discounts	8%	46	34	4.235
Other	18%	62	76.5	2.748
				9.501

$\chi^2 = 9.501$

(d) Reject H_0. There is enough evidence at the 5% significance level to conclude that the distribution of responses differs from the claimed or expected distribution.

9. (a) Claimed distribution:

Response	Distribution
Keeping Medicare, Social Security and Medicaid	77%
Tax cuts	15%
Not sure	8%

H_0: Distribution of responses is as shown in the table above.

H_a: Distribution of responses differs from the claimed or expected distribution. (claim)

(b) $\chi_0^2 = 9.210$; Reject H_0 if $\chi^2 > 9.210$.

(c)

Response	Distribution	Observed	Expected	$\frac{(O-E)^2}{E}$
Keeping Medicare, Social Security, and Medicaid	77%	431	462	2.080
Tax cuts	15%	107	90	3.211
Not sure	8%	62	48	4.083
				9.374

$\chi^2 = 9.374$

(d) Reject H_0. There is enough evidence at the 1% significance level to conclude that the distribution of responses differs from the claimed or expected distribution.

11. (a) Claimed distribution:

Day	Distribution
Sunday	14.286%
Monday	14.286%
Tuesday	14.286%
Wednesday	14.286%
Thursday	14.286%
Friday	14.286%
Saturday	14.286%

H_0: The distribution of fatal bicycle accidents throughout the week is uniform as shown in the table above. (claim)

H_a: The distribution of fatal bicycle accidents throughout the week is not uniform.

(b) $\chi_0^2 = 10.645$; Reject H_0 if $\chi^2 > 10.645$.

(c)

Day	Distribution	Observed	Expected	$\frac{(O - E)^2}{E}$
Sunday	14.286%	108	111.717	0.124
Monday	14.286%	112	111.717	0.0007
Tuesday	14.286%	105	111.717	0.404
Wednesday	14.286%	111	111.717	0.005
Thursday	14.286%	123	111.717	1.140
Friday	14.286%	105	111.717	0.404
Saturday	14.286%	118	111.717	0.353
				2.430

$\chi^2 \approx 2.430$

(d) Fail to reject H_0. There is not enough evidence at the 10% significance level to reject the claim that the distribution of fatal bicycle accidents throughout the week is uniform.

13. (a) Claimed distribution:

Object struck	Distribution
Tree	50%
Utility pole	14%
Embankment	6%
Guardrail	5%
Ditch	3%
Culvert	2%
Other	20%

H_0: Distribution of objects struck is as shown in table above.

H_a: Distribution of objects struck differs from the claimed or expected distribution. (claim)

(b) $\chi_0^2 = 16.812$; Reject H_0 if $\chi^2 > 16.812$.

(c)

Object Struck	Distribution	Observed	Expected	$\frac{(O - E)^2}{E}$
Tree	50%	375	345.50	2.519
Utility pole	14%	86	96.74	1.192
Embankment	6%	53	41.46	3.212
Guardrail	5%	42	34.55	1.606
Ditch	3%	27	20.73	1.896
Culvert	2%	18	13.82	1.264
Other	20%	90	138.2	16.811
			691	28.501

$\chi^2 \approx 28.501$

(d) Reject H_0. There is enough evidence at the 1% significance level to conclude that the distribution of objects struck changed from the claimed or expected distribution.

15. (a) Claimed distribution:

Response	Distribution
Not a HS grad	33.333%
HS graduate	33.333%
College (1yr+)	33.333%

H_0: Distribution of the responses is uniform as shown in table above. (claim)

H_a: Distribution of the responses is not uniform.

(b) $\chi_0^2 = 7.378$; Reject H_0 if $\chi^2 > 7.378$.

(c)

Response	Distribution	Observed	Expected	$\frac{(O - E)^2}{E}$
Not a HS grad	33.333%	37	33	0.485
HS graduate	33.333%	40	33	1.485
College (1yr+)	33.333%	22	33	3.667
			99	5.637

$\chi^2 \approx 5.637$

(d) Fail to reject H_0. There is not enough evidence at the 2.5% significance level to reject the claim that the distribution of the responses is uniform.

17. (a) Claimed distribution:

Cause	Distribution
Trans. Accidents	43%
Objects/equipment	18%
Assaults	14%
Falls	13%
Exposure	9%
Fire/explosions	3%

H_0: Distribution of the causes is as shown in table above.

H_a: Distribution of the causes differs from the claimed or expected distribution. (claim)

(b) $\chi_0^2 = 11.071$; Reject H_0 if $\chi^2 > 11.071$.

(c)

Cause	Distribution	Observed	Expected	$\frac{(O - E)^2}{E}$
Trans. Accidents	43%	2891	2679.33	16.722
Objects/equipment	18%	1159	1121.58	1.248
Assaults	14%	804	872.34	5.354
Falls	13%	754	810.03	3.876
Exposure	9%	531	560.79	1.582
Fires/explosions	3%	92	186.93	48.209
		6231		76.992

$\chi^2 \approx 76.992$

(d) Reject H_0. There is enough evidence at the 5% significance level to conclude that the distributions of the causes in the Western U.S. differs from the national distribution.

19. (a) Frequency distribution: $\mu = 69.435$; $\sigma \approx 8.337$

Lower Boundary	Upper Boundary	Lower z-score	Upper z-score	Area
49.5	58.5	−2.39	−1.31	0.0867
58.5	67.5	−1.31	−0.23	0.3139
67.5	76.5	−0.23	0.85	0.3933
76.5	85.5	0.85	1.93	0.1709
85.5	94.5	1.93	3.01	0.0255

Class Boundaries	Distribution	Frequency	Expected	$\frac{(O-E)^2}{E}$
49.5–58.5	8.67%	19	17	0.235
58.5–67.5	31.39%	61	63	0.063
67.5–76.5	39.33%	82	79	0.114
76.5–85.5	17.09%	34	34	0
85.5–94.5	2.55%	4	5	0.2
		200		0.612

H_0: Test scores have a normal distribution. (claim)

H_a: Test scores do not have a normal distribution.

(b) $\chi_0^2 = 13.277$; Reject H_0 if $\chi^2 > 13.277$.

(c) $\chi^2 = 0.612$

(d) Fail to reject H_0. There is not enough evidence at the 1% significance level to reject the claim that the distribution of test scores is normal.

10.2 INDEPENDENCE

10.2 Try It Yourself Solutions

1ab.

	Hotel	Leg Room	Rental Size	Other	Total
Business	36	108	14	22	180
Leisure	38	54	14	14	120
Total	74	162	28	36	300

c. $n = 300$

d.

	Hotel	Leg Room	Rental Size	Other
Business	44.4	97.2	16.8	21.6
Leisure	29.6	64.8	11.2	14.4

2a. H_0: Travel concern is independent of travel purpose.

H_a: Travel concern is dependent on travel purpose. (claim)

b. $\alpha = 0.01$ **c.** $(r-1)(c-1) = 3$ **d.** $\chi_0^2 = 11.345$; Reject H_0 if $\chi^2 > 11.345$.

e. $\chi^2 = \sum \frac{(O-E)^2}{E} = \frac{(36-44.4)^2}{44.4} + \frac{(108-97.2)^2}{97.2} + \cdots + \frac{(14-14.4)^2}{14.4} \approx 8.158$

f. Fail to reject H_0.

g. There is not enough evidence to conclude that travel concern is dependent on travel purpose.

3a. H_0: The number of minutes adults spend online per day is independent on gender.

H_a: The number of minutes adults spend online per day is dependent on gender. (claim)

b. Enter the data.   **c.** $\chi_0^2 = 9.488$; Reject H_0 if $\chi^2 > 9.488$.

d. $\chi^2 \approx 65.619$ **e.** Reject H_0.

f. There is enough evidence to conclude that minutes spent online per day is dependent on gender.

10.2 EXERCISE SOLUTIONS

1. $E_{r,c} = \dfrac{(\text{Sum of row } r)(\text{Sum of column } c)}{\text{Sample size}}$

3. False. In order to use the χ^2 independence test, each expected frequency must be greater than or equal to 5.

5. False. If the two variables of a chi-square test for independence are dependent, then you can expect to find large differences between the observed frequencies and the expected frequencies.

7. (a)

	Athlete has		
Result	Stretched	Not stretched	
Injury	18	22	40
No injury	211	189	400
	229	211	440

(b)

	Athlete has	
Result	Stretched	Not stretched
Injury	20.818	19.182
No injury	208.182	191.818

9. (a)

	Treatment			
Result	Brand-name	Generic	Placebo	
Improvement	24	21	10	55
No change	12	13	45	70
	36	34	55	125

(b)

	Treatment		
Result	Brand-name	Generic	Placebo
Improvement	15.84	14.96	24.2
No change	20.16	19.04	30.8

11. (a)

	Type of Car				
Gender	Compact	Full-size	SUV	Truck/Van	
Male	28	39	21	22	110
Female	24	32	20	14	90
	52	71	41	36	200

(b)

	Type of Car			
Gender	Compact	Full-size	SUV	Truck/Van
Male	28.60	39.05	22.55	19.80
Female	23.40	31.95	18.45	16.20

13. (a) H_0: Skill level in a subject is independent of location. (claim)

    H_a: Skill level in a subject is dependent on location.

    (b) d.f. $= (r - 1)(c - 1) = 2$

    $\chi_0^2 = 9.210$; Reject H_0 if $\chi^2 > 9.210$.

    (c) $\chi^2 \approx 0.297$

    (d) Fail to reject H_0. There is not enough evidence at the 1% significance level to reject the claim that skill level in a subject is independent of location.

15. (a) H_0: Grades are independent of the institution.

    H_a: Grades are dependent on the institution. (claim)

    (b) d.f. $= (r - 1)(c - 1) = 8$

    $\chi_0^2 = 15.507$; Reject H_0 if $\chi^2 > 15.507$.

    (c) $\chi^2 \approx 48.488$

    (d) Reject H_0. There is enough evidence at the 5% significance level to conclude that grades are dependent on the institution.

17. (a) H_0: Results are independent of the type of treatment.

    H_a: Results are dependent on the type of treatment. (claim)

    (b) d.f. $= (r - 1)(c - 1) = 1$

    $\chi_0^2 = 2.706$; Reject H_0 if $\chi^2 > 2.706$.

    (c) $\chi^2 \approx 5.106$

    (d) Reject H_0. There is enough evidence at the 10% significance level to conclude that results are dependent on the type of treatment. I would recommend using the drug.

19. (a) H_0: Reasons are independent of the type of worker.

    H_a: Reasons are dependent on the type of worker. (claim)

    (b) d.f. $= (r - 1)(c - 1) = 2$

    $\chi_0^2 = 9.210$; Reject H_0 if $\chi^2 > 9.210$.

    (c) $\chi^2 \approx 7.326$

    (d) Fail to reject H_0. There is not enough evidence at the 1% significance level to conclude that the reason(s) for continuing education are dependent on the type of worker. Based on these results, marketing strategies should not differ between technical and non-technical audiences in regard to reason(s) for continuing education.

21. (a) H_0: Type of crash is independent of the type of vehicle.

   H_a: Type of crash is dependent on the type of vehicle. (claim)

(b) d.f. $= (r - 1)(c - 1) = 2$

   $\chi_0^2 = 5.991$; Reject H_0 if $\chi^2 > 5.991$.

(c) $\chi^2 \approx 108.913$

(d) Reject H_0. There is enough evidence at the 5% significance level to conclude that type of crash is dependent on the type of vehicle.

23. (a) H_0: Subject is independent of coauthorship.

   H_a: Subject and coauthorship are dependent. (claim)

(b) d.f. $= (r - 1)(c - 1) = 4$

   $\chi_0^2 = 7.779$; Reject H_0 if $\chi^2 > 7.779$.

(c) $\chi^2 \approx 5.610$

(d) Fail to reject H_0. There is not enough evidence at the 10% significance level to conclude that the subject matter and coauthorship are related.

25. H_0: The proportions are equal. (claim)

   H_a: At least one of the proportions is different from the others.

   d.f. $= (r - 1)(c - 1) = 7$

   $\chi_0^2 = 14.067$; Reject H_0 if $\chi^2 > 14.067$.

   $\chi^2 \approx 7.462$

   Fail to reject H_0. There is not enough evidence at the 5% significance level to reject the claim that the proportions are equal.

27.

Status	Educational Attainment			
	Not a high school graduate	High school graduate	Some college, no degree	Associate's, Bachelor's, or advanced degree
Employed	0.0610	0.1903	0.1129	0.2725
Unemployed	0.0058	0.0111	0.0053	0.0074
Not in labor force	0.0817	0.1204	0.0498	0.0817

29. 8.2%   **31.** 63.7%

33. Several of the expected frequencies are less than 5.

35. 29.9%

37.

Status	Educational Attainment			
	Not a high school graduate	High school graduate	Some college, no degree	Associate's, Bachelor's, or advanced degree
Employed	0.4107	0.5914	0.6719	0.7537
Unemployed	0.0393	0.0346	0.0315	0.0205
Not in labor force	0.5500	0.3740	0.2965	0.2258

39. 3.9%

41. As educational attainment increases, employment increases.

10.3 COMPARING TWO VARIANCES

10.3 Try It Yourself Solutions

1a. $\alpha = 0.01$

  b. $F = 5.42$

2a. $\alpha = 0.01$

  b. $F = 18.31$

3a. $H_0: \sigma_1^2 \leq \sigma_2^2; H_a: \sigma_1^2 > \sigma_2^2$ (claim)

  b. $\alpha = 0.01$

  c. d.f.$_N = n_1 - 1 = 24$
     d.f.$_D = n_2 - 1 = 19$

  d. $F_0 = 2.92$; Reject H_0 if $F > 2.92$.

  e. $F = \dfrac{s_1^2}{s_2^2} = \dfrac{180}{56} \approx 3.21$

  f. Reject H_0.

  g. There is enough evidence to support the claim.

4a. $H_0: \sigma_1 = \sigma_2$ (claim); $H_a: \sigma_1 \neq \sigma_2$

  b. $\alpha = 0.01$

  c. d.f.$_N = n_1 - 1 = 15$
     d.f.$_D = n_2 - 1 = 21$

  d. $F_0 = 3.43$; Reject H_0 if $F > 3.43$.

  e. $F = \dfrac{s_1^2}{s_2^2} = \dfrac{(0.95)^2}{(0.78)^2} \approx 1.48$

  f. Fail to reject H_0.

  g. There is not enough evidence to reject the claim.

10.3 EXERCISE SOLUTIONS

1. Specify the level of significance α. Determine the degrees of freedom for the numerator and denominator. Use Table 7 in Appendix B to find the critical value F.

3. (1) The samples must be randomly selected, (2) The samples must be independent, and (3) Each population must have a normal distribution.

5. $F = 2.54$

7. $F = 4.86$

9. $F = 2.06$

11. $H_0: \sigma_1^2 \le \sigma_2^2; H_a: \sigma_1^2 > \sigma_2^2$ (claim)

d.f.$_N$ = 4

d.f.$_D$ = 5

$F_0 = 3.52$; Reject H_0 if $F > 3.52$.

$F = \dfrac{s_1^2}{s_2^2} = \dfrac{773}{765} \approx 1.010$

Fail to reject H_0. There is not enough evidence to support the claim.

13. $H_0: \sigma_1^2 \le \sigma_2^2$ (claim); $H_a: \sigma_1^2 > \sigma_2^2$

d.f.$_N$ = 10

d.f.$_D$ = 9

$F_0 = 5.26$; Reject H_0 if $F > 5.26$.

$F = \dfrac{s_1^2}{s_2^2} = \dfrac{842}{836} \approx 1.007$

Fail to reject H_0. There is not enough evidence to reject the claim.

15. $H_0: \sigma_1^2 = \sigma_2^2$ (claim); $H_a: \sigma_1^2 \ne \sigma_2^2$

d.f.$_N$ = 12

d.f.$_D$ = 19

$F_0 = 3.30$; Reject H_0 if $F > 3.30$.

$F = \dfrac{s_1^2}{s_2^2} = \dfrac{9.8}{2.5} \approx 3.920$

Reject H_0. There is enough evidence to reject the claim.

17. Population 1: Company B

Population 2: Company A

(a) $H_0: \sigma_1^2 \le \sigma_2^2; H_a: \sigma_1^2 > \sigma_2^2$ (claim)

(b) d.f.$_N$ = 24

   d.f.$_D$ = 19

   $F_0 = 2.11$; Reject H_0 if $F > 2.11$.

(c) $F = \dfrac{s_1^2}{s_2^2} = \dfrac{2.8}{2.6} \approx 1.08$

(d) Fail to reject H_0.

(e) There is not enough evidence at the 5% significance level to support Company A's claim that the variance of life of its appliances is less than the variance of life of Company B appliances.

19. Population 1: District 1

    Population 2: District 2

    (a) H_0: $\sigma_1^2 = \sigma_2^2$ (claim); H_a: $\sigma_1^2 \neq \sigma_2^2$

    (b) d.f.$_N$ = 11

      d.f.$_D$ = 13

      $F_0 = 2.635$; Reject H_0 if $F > 2.635$.

    (c) $F = \dfrac{s_1^2}{s_2^2} = \dfrac{(36.8)^2}{(32.5)^2} \approx 1.282$

    (d) Fail to reject H_0.

    (e) There is not enough evidence at the 10% significance level to reject the claim that the standard deviation of science assessment test scores for eighth grade students is the same in Districts 1 and 2.

21. Population 1: Before new admissions procedure

    Population 2: After new admissions procedure

    (a) H_0: $\sigma_1^2 \leq \sigma_2^2$; H_a: $\sigma_1^2 > \sigma_2^2$ (claim)

    (b) d.f.$_N$ = 24

      d.f.$_D$ = 20

      $F_0 = 1.77$; Reject H_0 if $F > 1.77$.

    (c) $F = \dfrac{s_1^2}{s_2^2} = \dfrac{(0.7)^2}{(0.5)^2} \approx 1.96$

    (d) Reject H_0.

    (e) There is enough evidence at the 10% significance level to support the claim that the standard deviation of waiting times has decreased.

23. (a) Population 1: New York

      Population 2: California

      H_0: $\sigma_1^2 \leq \sigma_2^2$; H_a: $\sigma_1^2 > \sigma_2^2$ (claim)

    (b) d.f.$_N$ = 15

      d.f.$_D$ = 16

      $F_0 = 2.35$; Reject H_0 if $F > 2.35$.

    (c) $F = \dfrac{s_1^2}{s_2^2} = \dfrac{(14,900)^2}{(9,600)^2} \approx 2.41$

    (d) Reject H_0.

    (e) There is enough evidence at the 5% significance level to conclude the standard deviation of annual salaries for actuaries is greater in New York than in California.

25. Right-tailed: $F_R = 14.73$

Left-tailed:

(1) d.f.$_N$ = 3 and d.f.$_D$ = 6

(2) $F = 6.60$

(3) Critical value is $\dfrac{1}{F} = \dfrac{1}{6.60} \approx 0.15$.

27. $\dfrac{s_1^2}{s_2^2} F_L < \dfrac{\sigma_1^2}{\sigma_2^2} < \dfrac{s_1^2}{s_2^2} F_R \rightarrow \dfrac{10.89}{9.61}\,0.331 < \dfrac{\sigma_1^2}{\sigma_2^2} < \dfrac{10.89}{9.61}\,3.33 \rightarrow 0.375 < \dfrac{\sigma_1^2}{\sigma_2^2} < 3.774$

10.4 ANALYSIS OF VARIANCE

10.4 Try It Yourself Solutions

1a. H_0: $\mu_1 = \mu_2 = \mu_3 = \mu_4$

H_a: At least one mean is different from the others. (claim)

b. $\alpha = 0.05$

c. d.f.$_N$ = 3

d.f.$_D$ = 14

d. $F_0 = 3.34$; Reject H_0 if $F > 3.34$.

e.

Variation	Sum of Squares	Degrees of Freedom	Mean Squares	F
Between	549.8	3	183.3	4.22
Within	608.0	14	43.4	

$F \approx 4.22$

f. Reject H_0.

g. There is enough evidence to conclude that at least one mean is different from the others.

2a. Enter the data.

b. H_0: $\mu_1 = \mu_2 = \mu_3 = \mu_4$

H_a: At least one mean is different from the others. (claim)

Variation	Sum of Squares	Degrees of Freedom	Mean Squares	F
Between	0.584	3	0.195	1.34
Within	4.360	30	0.145	

$F = 1.34 \rightarrow P\text{-value} = 0.280$

c. $0.280 > 0.05$

d. Fail to reject H_0. There is not enough evidence to conclude that at least one mean is different from the others.

10.4 EXERCISE SOLUTIONS

1. $H_0: \mu_1 = \mu_2 = \ldots = \mu_k$

$H_a:$ At least one of the means is different from the others.

3. MS_B measures the differences related to the treatment given to each sample.

MS_W measures the differences related to entries within the same sample.

5. (a) $H_0: \mu_1 = \mu_2 = \mu_3$

$H_a:$ At least one mean is different from the others. (claim)

(b) $\text{d.f.}_N = k - 1 = 2$

$\text{d.f.}_D = N - k = 26$

$F_0 = 3.37$; Reject H_0 if $F > 3.37$.

(c)

Variation	Sum of Squares	Degrees of Freedom	Mean Squares	F
Between	0.518	2	0.259	1.017
Within	6.629	26	0.255	

$F \approx 1.02$

(d) Fail to reject H_0. There is not enough evidence at the 5% significance level to conclude that the mean costs per ounce are different.

7. (a) $H_0: \mu_1 = \mu_2 = \mu_3$ (claim)

$H_a:$ At least one mean is different from the others.

(b) $\text{d.f.}_N = k - 1 = 2$

$\text{d.f.}_D = N - k = 12$

$F_0 = 2.81$; Reject H_0 if $F > 2.81$.

(c)

Variation	Sum of Squares	Degrees of Freedom	Mean Squares	F
Between	302.6	2	151.3	1.77
Within	1024.2	12	85.4	

$F \approx 1.77$

(d) Fail to reject H_0. There is not enough evidence at the 10% significance level to reject the claim that the mean prices are all the same for the three types of treatment.

9. (a) $H_0: \mu_1 = \mu_2 = \mu_3 = \mu_4$ (claim)

$H_a:$ At least one mean is different from the others.

(b) $\text{d.f.}_N = k - 1 = 3$

$\text{d.f.}_D = N - k = 29$

$F_0 = 4.54$; Reject H_0 if $F > 4.54$.

(c)

Variation	Sum of Squares	Degrees of Freedom	Mean Squares	F
Between	5.608	3	1.869	0.557
Within	97.302	29	3.355	

$F \approx 0.56$

(d) Fail to reject H_0. There is not enough evidence at the 1% significance level to reject the claim that the mean number of days patients spend in the hospital is the same for all four regions.

11. (a) H_0: $\mu_1 = \mu_2 = \mu_3 = \mu_4$ (claim)

   H_a: At least one mean is different from the others.

   (b) d.f.$_N = k - 1 = 3$

   d.f.$_D = N - k = 40$

   $F_0 = 2.23$; Reject H_0 if $F > 2.23$.

   (c)

Variation	Sum of Squares	Degrees of Freedom	Mean Squares	F
Between	15,095.256	3	5031.752	2.757
Within	73,015.903	43	1825.398	

   $F \approx 2.76$

   (d) Reject H_0. There is enough evidence at the 10% significance level to reject the claim that the mean price is the same for all four cities.

13. (a) H_0: $\mu_1 = \mu_2 = \mu_3 = \mu_4$

   H_a: At least one mean is different from the others. (claim)

   (b) d.f.$_N = k - 1 = 3$

   d.f.$_D = N - k = 26$

   $F_0 = 2.98$; Reject H_0 if $F > 2.98$

   (c)

Variation	Sum of Squares	Degrees of Freedom	Mean Squares	F
Between	2,220,266,803	3	740,088,934	16.195
Within	1,188,146,984	16	45,697,960.9	

   $F \approx 16.20$

   (d) Reject H_0. There is enough evidence at the 10% significance level to conclude that at least one of the mean prices is different from the others.

15. (a) H_0: $\mu_1 = \mu_2 = \mu_3 = \mu_4$ (claim)

   H_a: At least one mean is different from the others.

   (b) d.f.$_N = k - 1 = 3$

   d.f.$_D = N - k = 28$

   $F_0 = 4.57$; Reject H_0 if $F > 4.57$.

   (c)

Variation	Sum of Squares	Degrees of Freedom	Mean Squares	F
Between	771.25	3	257.083	0.459
Within	15,674.75	28	559.813	

   $F \approx 0.46$

   (d) Fail to reject H_0. There is not enough evidence at the 1% significance level to reject the claim that the mean numbers of female students are equal for all grades.

17. H_0: Advertising medium has no effect on mean ratings.

H_a: Advertising medium has an effect on mean ratings.

H_0: Length of ad has no effect on mean ratings.

H_a: Length of ad has an effect on mean ratings.

H_0: There is no interaction effect between advertising medium and length of ad on mean ratings.

H_a: There is an interaction effect between advertising medium and length of ad on mean ratings.

Source	d.f.	SS	MS	F	P
Ad medium	1	1.25	1.25	0.57	0.459
Length of ad	1	0.45	0.45	0.21	0.655
Interaction	1	0.45	0.45	0.21	0.655
Error	16	34.80	2.17		
Total	19	36.95			

None of the null hypotheses can be rejected at the 10% significance level.

19. H_0: Age has no effect on mean GPA.

H_a: Age has an effect on mean GPA.

H_0: Gender has no effect on mean GPA.

H_a: Gender has an effect on mean GPA.

H_0: There is no interaction effect between age and gender on mean GPA.

H_a: There is an interaction effect between age and gender on mean GPA.

Source	d.f.	SS	MS	F	P
Age	3	0.41	0.14	0.12	0.948
Gender	1	0.18	0.18	0.16	0.697
Interaction	3	0.29	0.10	0.08	0.968
Error	16	18.66	1.17		
Total	23	19.55			

None of the null hypotheses can be rejected at the 10% significance level.

21.

	Mean	Size
Pop 1	17.82	13
Pop 2	13.50	13
Pop 3	13.12	14
Pop 4	12.91	14

$SS_w = 977.796$

$\Sigma(n_i - 1) = N - k = 50$

$F_0 = 2.205 \rightarrow CV_{\text{Scheffé}} = 2.205(4 - 1) = 6.615$

$$\frac{(\bar{x}_1 - \bar{x}_2)^2}{\dfrac{SS_w}{\Sigma(n_i - 1)}\left[\dfrac{1}{n_1} + \dfrac{1}{n_2}\right]} \approx 6.212 \rightarrow \text{No difference}$$

$$\frac{(\bar{x}_1 - \bar{x}_3)^2}{\dfrac{SS_w}{\Sigma(n_i - 1)}\left[\dfrac{1}{n_1} + \dfrac{1}{n_3}\right]} \approx 7.620 \rightarrow \text{Significant difference}$$

$$\frac{(\bar{x}_1 - \bar{x}_4)^2}{\dfrac{SS_w}{\Sigma(n_i - 1)}\left[\dfrac{1}{n_1} + \dfrac{1}{n_4}\right]} \approx 8.330 \rightarrow \text{Significant difference}$$

$$\frac{(\bar{x}_2 - \bar{x}_3)^2}{\dfrac{SS_w}{\Sigma(n_i - 1)}\left[\dfrac{1}{n_2} + \dfrac{1}{n_3}\right]} \approx 0.049 \rightarrow \text{No difference}$$

$$\frac{(\bar{x}_2 - \bar{x}_4)^2}{\dfrac{SS_w}{\Sigma(n_i - 1)}\left[\dfrac{1}{n_2} + \dfrac{1}{n_4}\right]} \approx 0.121 \rightarrow \text{No difference}$$

$$\frac{(\bar{x}_3 - \bar{x}_4)^2}{\dfrac{SS_w}{\Sigma(n_i - 1)}\left[\dfrac{1}{n_3} + \dfrac{1}{n_4}\right]} \approx 0.016 \rightarrow \text{No difference}$$

23.

	Mean	Size
Pop 1	66,492	8
Pop 2	60,528	7
Pop 3	55,504	9
Pop 4	79,500	6

$SS_w \approx 1{,}188{,}146{,}984$

$\Sigma(n_i - 1) = N - k = 26$

$F_0 = 2.98 \rightarrow CV_{\text{Scheffé}} = 2.98(4 - 1) = 8.94$

$$\frac{(\bar{x}_1 - \bar{x}_2)^2}{\dfrac{SS_w}{\Sigma(n_i - 1)}\left[\dfrac{1}{n_1} + \dfrac{1}{n_2}\right]} \approx 2.91 \rightarrow \text{No difference}$$

$$\frac{(\bar{x}_1 - \bar{x}_3)^2}{\dfrac{SS_w}{\Sigma(n_i - 1)}\left[\dfrac{1}{n_1} + \dfrac{1}{n_3}\right]} \approx 11.19 \rightarrow \text{Significant difference}$$

$$\frac{(\bar{x}_1 - \bar{x}_4)^2}{\dfrac{SS_w}{\Sigma(n_i - 1)}\left[\dfrac{1}{n_1} + \dfrac{1}{n_4}\right]} \approx 12.70 \rightarrow \text{Significant difference}$$

$$\frac{(\bar{x}_2 - \bar{x}_3)^2}{\dfrac{SS_w}{\Sigma(n_i - 1)}\left[\dfrac{1}{n_2} + \dfrac{1}{n_3}\right]} \approx 2.17 \rightarrow \text{No difference}$$

$$\frac{(\bar{x}_2 - \bar{x}_4)^2}{\dfrac{SS_w}{\Sigma(n_i - 1)}\left[\dfrac{1}{n_2} + \dfrac{1}{n_4}\right]} \approx 25.45 \rightarrow \text{Significant difference}$$

$$\frac{(\bar{x}_3 - \bar{x}_4)^2}{\dfrac{SS_w}{\Sigma(n_i - 1)}\left[\dfrac{1}{n_3} + \dfrac{1}{n_4}\right]} \approx 45.36 \rightarrow \text{Significant difference}$$

CHAPTER 10 REVIEW EXERCISE SOLUTIONS

1. Claimed distribution:

Category	Distribution
0	16%
1–3	46%
4–9	25%
10+	13%

H_0: Distribution of health care visits is as shown in table above.

H_a: Distribution of health care visits differs from the claimed distribution.

$\chi_0^2 = 7.815$

Category	Distribution	Observed	Expected	$\frac{(O - E)^2}{E}$
0	16%	99	117.44	2.895
1–3	46%	376	337.64	4.368
4–9	25%	167	183.50	1.484
10+	13%	92	95.42	0.123
		734		8.860

$\chi_0^2 = 8.860$

Reject H_0. There is enough evidence at the 5% significance level to reject the claimed or expected distribution.

3. Claimed distribution:

Days	Distribution
0	19%
1–5	30%
6–10	30%
11–15	12%
16–20	3%
21+	6%

H_0: Distribution of days is as shown in the table above.

H_a: Distribution of days differs from the claimed distribution.

$\chi_0^2 = 9.236$

Days	Distribution	Observed	Expected	$\frac{(O - E)^2}{E}$
0	19%	165	152	1.112
1–5	30%	237	240	0.038
6–10	30%	245	240	0.104
11–15	12%	88	96	0.667
16–20	3%	19	24	1.042
21+	6%	46	48	0.083
		800		3.045

$\chi^2 \approx 3.045$

Fail to reject H_0. There is not enough evidence at the 10% significance level to support the claim.

5. (a) Expected frequencies:

	HS–did not complete	HS complete	College 1–3 years	College 4+ years	Total
25–44	663.49	1440.47	1137.54	1237.5	4479
45+	856.51	1859.53	1468.46	1597.5	5782
Total	1520	3300	2606	2835	10,261

(b) H_0: Education is independent of age.

H_a: Education is dependent on age.

d.f. = 3

$\chi_0^2 = 6.251$

$\chi^2 = \sum \dfrac{(O - E)^2}{E}$ 66.128

Reject H_0.

(c) There is enough evidence at the 10% significance level to conclude that education level of people in the United States and their age are dependent.

7. (a) Expected frequencies:

Gender	Age Group						Total
	16–20	21–30	31–40	41–50	51–60	61+	
Male	143.76	284.17	274.14	237.37	147.10	43.46	1130
Female	71.24	140.83	135.86	117.63	72.90	21.54	560
Total	215	425	410	355	220	65	1690

(b) H_0: Gender is independent of age.

H_a: Gender and age are dependent.

d.f. = 5

$\chi_0^2 = 11.071$

$\chi^2 = \sum \dfrac{(O - E)^2}{E} = 9.951$

Fail to reject H_0.

(c) There is not enough evidence at the 10% significance level to conclude that gender and age group are dependent.

9. $F_0 \approx 2.295$ **11.** $F_0 = 2.39$

13. H_0: $\sigma_1^2 \leq \sigma_2^2$ (claim); H_a: $\sigma_1^2 > \sigma_2^2$

d.f.$_N$ = 15

d.f.$_D$ = 20

$F_0 = 3.09$; Reject H_0 if $F > 3.09$.

$F = \dfrac{s_1^2}{s_2^2} = \dfrac{653}{270} \approx 2.419$

Fail to reject H_0. There is not enough evidence to reject the claim.

15. Population 1: Garfield County

Population 2: Kay County

$H_0: \sigma_1^2 \le \sigma_2^2; H_a: \sigma_1^2 > \sigma_2^2$ (claim)

d.f.$_N$ = 20

d.f.$_D$ = 15

$F_0 = 1.92$; Reject H_0 if $F > 1.92$.

$F = \dfrac{s_1^2}{s_2^2} = \dfrac{(0.76)^2}{(0.58)^2} \approx 1.717$

Fail to reject H_0. There is not enough evidence at the 10% significance level to support the claim that the variation in wheat production is greater in Garfield County than in Kay County.

17. Population 1: Male $\rightarrow s_1^2 = 18{,}486.26$

Population 2: Female $\rightarrow s_2^2 = 12{,}102.78$

$H_0: \sigma_1^2 = \sigma_2^2; H_a: \sigma_1^2 \ne \sigma_2^2$ (claim)

d.f.$_N$ = 13

d.f.$_D$ = 8

$F_0 = 6.94$; Reject H_0 if $F > 6.94$.

$F = \dfrac{s_1^2}{s_2^2} = \dfrac{18{,}486.26}{12{,}102.78} \approx 1.527$

Fail to reject H_0. There is not enough evidence at the 1% significance level to support the claim that the test score variance for females is different than that for males.

19. $H_0: \mu_1 = \mu_2 = \mu_3 = \mu_4$

H_a: At least one mean is different from the others. (claim)

d.f.$_N$ = $k - 1$ = 3

d.f.$_D$ = $N - k$ = 28

$F_0 = 2.29$; Reject H_0 if $F > 2.29$.

Variation	Sum of Squares	Degrees of Freedom	Mean Squares	F
Between	512.457	3	170.819	8.508
Within	562.162	28	20.077	

$F \approx 8.508$

Reject H_0. There is enough evidence at the 10% significance level to conclude that the mean residential energy expenditures are not the same for all four regions.

CHAPTER 10 QUIZ SOLUTIONS

1. (a) Population 1: San Jose $\rightarrow s_1^2 \approx 430.084$

Population 2: Dallas $\rightarrow s_2^2 \approx 120.409$

$H_0: \sigma_1^2 = \sigma_2^2; H_a: \sigma_1^2 \ne \sigma_2^2$ (claim)

(b) $\alpha = 0.01$

(cd) $\text{d.f.}_N = 12$

$\text{d.f.}_D = 15$

$F_0 = 4.25$; Reject H_0 if $F > 4.25$.

(e) $F = \dfrac{s_1^2}{s_2^2} = \dfrac{430.084}{120.409} \approx 3.57$

(f) Fail to reject H_0.

(g) There is not enough evidence at the 1% significance level to conclude that the variances in annual wages for San Jose, CA and Dallas, TX are different.

2. (a) $H_0\colon \mu_1 = \mu_2 = \mu_3$ (claim)

$H_a\colon$ At least one mean is different from the others.

(b) $\alpha = 0.10$

(cd) $\text{d.f.}_N = k - 1 = 2$

$\text{d.f.}_D = N - k = 40$

$F_0 = 2.44$; Reject H_0 if $F > 2.44$.

Variation	Sum of Squares	Degrees of Freedom	Mean Squares	F
Between	5021.896	2	2510.948	12.213
Within	8224.121	40	205.603	

(e) $F \approx 12.21$

(f) Reject H_0.

(g) There is enough evidence at the 10% significance level to reject the claim that the mean annual wages for the three cities are not all equal.

3. (a) Claimed distribution:

Education	25 & Over
Not a HS graduate	14.8%
HS graduate	32.2%
Some college, no degree	16.8%
Associate's degree	8.6%
Bachelor's degree	18.1%
Advanced degree	9.5%

$H_0\colon$ Distribution of educational achievement for people in the United States ages 35–44 is as shown in table above.

$H_a\colon$ Distribution of educational achievement for people in the United States ages 35–44 differs from the claimed distribution. (claim)

(b) $\alpha = 0.01$

(cd) $\chi_0^2 = 15.086$; Reject H_0 if $\chi^2 > 15.086$.

(e)

Education	25 & Over	Observed	Expected	$\frac{(O-E)^2}{E}$
Not a HS graduate	14.8%	35	44.548	2.046
HS graduate	32.2%	95	96.922	0.038
Some college, no degree	16.8%	51	50.568	0.004
Associate's degree	8.6%	30	25.886	0.654
Bachelor's degree	18.1%	61	54.481	0.780
Advanced degree	9.5%	29	28.595	0.006
		301		3.528

$\chi^2 = 3.528$

(f) Fail to reject H_0.

(g) There is not enough evidence at the 1% significance level to conclude that the distribution of educational achievement for people in the United States ages 35–44 differs from the claimed or expected distribution.

4. (a) Claimed distribution:

Education	25 & Over
Not a HS graduate	14.8%
HS graduate	32.2%
Some college, no degree	16.8%
Associate's degree	8.6%
Bachelor's degree	18.1%
Advanced degree	9.5%

H_0: Distribution of educational achievement for people in the United States ages 65–74 is as shown in table above.

H_a: Distribution of educational achievement for people in the United States ages 65–74 differs from the claimed distribution. (claim)

(b) $\alpha = 0.05$

(c) $\chi_0^2 = 11.071$

(d) Reject H_0 if $\chi^2 > 11.071$.

(e)

Education	25 & Over	Observed	Expected	$\frac{(O-E)^2}{E}$
Not a HS graduate	14.8%	91	60.088	15.903
HS graduate	32.2%	151	130.730	3.142
Some college, no degree	16.8%	58	68.208	1.528
Associate's degree	8.6%	23	34.916	4.067
Bachelor's degree	18.1%	50	73.486	7.506
Advanced degree	9.5%	33	38.570	0.804
		406		32.950

$\chi^2 = 32.950$

(f) Reject H_0.

(g) There is enough evidence at the 5% significance level to conclude that the distribution of educational achievement for people in the United States ages 65–74 differs from the claimed or expected distribution.

Nonparametric Tests

11.1 Try It Yourself Solutions

1a. H_0: median ≤ 2500; H_a: median > 2500 (claim)

b. $\alpha = 0.025$

c. $n = 22$

d. The critical value is 5.

e. $x = 10$

f. Fail to reject H_0.

g. There is not enough evidence to support the claim.

2a. H_0: median $= \$134{,}500$ (claim); H_a: median $\neq \$134{,}500$

b. $\alpha = 0.10$

c. $n = 81$

d. The critical value is $z_0 = -1.645$.

e. $x = 30$

$$z = \frac{(x + 0.5) - 0.5(n)}{\frac{\sqrt{n}}{2}} = \frac{(30 + 0.5) - 0.5(81)}{\frac{\sqrt{81}}{2}} = \frac{-10}{4.5} = -2.22$$

f. Reject H_0.

g. There is enough evidence to reject the claim.

3a. H_0: The number of colds will not decrease.
    H_a: The number of colds will decrease. (claim)

b. $\alpha = 0.05$

c. $n = 11$

d. The critical value is 2.

e. $x = 2$

f. Reject H_0.

g. There is enough evidence to support the claim.

11.1 EXERCISE SOLUTIONS

1. A nonparametric test is a hypothesis test that does not require any specific conditions concerning the shape of populations or the value of any population parameters.

A nonparametric test is usually easier to perform than its corresponding parametric test, but the nonparametric test is usually less efficient.

203

3. (a) H_0: median $\leq$ \$300; H_a: median $>$ \$300 (claim)

   (b) Critical value is 1.

   (c) $x = 5$

   (d) Fail to reject H_0.

   (e) There is not enough evidence at the 1% significance level to support the claim that the median amount of new credit card charges for the previous month was more than \$300.

5. (a) H_0: median $\leq$ \$210,000 (claim); H_a: median $>$ \$210,000

   (b) Critical value is 1.

   (c) $x = 3$

   (d) Fail to reject H_0.

   (e) There is not enough evidence at the 5% significance level to reject the claim that the median sales price of new privately owned one-family homes sold in the past year is \$210,000 or less.

7. (a) H_0: median $\geq$ \$2200 (claim); H_a: median $<$ \$2200

   (b) Critical value is $z_0 = -2.05$.

   (c) $x = 44$
   $$z = \frac{(x + 0.5) - 0.5(n)}{\frac{\sqrt{n}}{2}} = \frac{(44 + 0.5) - 0.5(104)}{\frac{\sqrt{104}}{2}} = \frac{-7.5}{5.099} \approx -1.47$$

   (d) Fail to reject H_0.

   (e) There is not enough evidence at the 2% significance level to reject the claim that the median amount of credit card debt for families holding such debts is at least \$2200.

9. (a) H_0: median $\leq$ 30; H_a: median $>$ 30 (claim)

   (b) Critical value is 2.

   (c) $x = 4$

   (d) Fail to reject H_0.

   (e) There is not enough evidence at the 1% significance level to support the claim that the median age of recipients of engineering doctorates is greater than 30 years.

11. (a) H_0: median $=$ 4 (claim); H_a: median $\neq$ 4

   (b) Critical value is $z_0 = -1.96$.

   (c) $x = 13$
   $$z = \frac{(x + 0.5) - 0.5(n)}{\frac{\sqrt{n}}{2}} = \frac{(13 + 0.5) - 0.5(33)}{\frac{\sqrt{33}}{2}} = \frac{-3}{2.872} \approx -1.04$$

   (d) Fail to reject H_0.

   (e) There is not enough evidence at the 5% significance level to reject the claim that the median number of rooms in renter-occupied units is 4.

13. (a) H_0: median = \$12.16 (claim); H_a: median ≠ \$12.16

   (b) Critical value is $z_0 = -2.575$.

   (c) $x = 16$

   $$z = \frac{(x + 0.5) - 0.5(n)}{\dfrac{\sqrt{n}}{2}} = \frac{(16 + 0.5) - 0.5(39)}{\dfrac{\sqrt{39}}{2}} = \frac{-3}{3.1225} \approx -0.961$$

   (d) Fail to reject H_0.

   (e) There is not enough evidence at the 1% significance level to reject the claim that the median hourly earnings of male workers paid hourly rates is \$12.16.

15. (a) H_0: The lower back pain intensity scores have not decreased.

   H_a: The lower back pain intensity scores have decreased. (claim)

   (b) Critical value is 1.

   (c) $x = 0$

   (d) Reject H_0.

   (e) There is enough evidence at the 5% significance level to conclude that the lower back pain intensity scores were reduced after accupunture.

17. (a) H_0: The SAT scores have not improved.

   H_a: The SAT scores have improved. (claim)

   (b) Critical value is 2.

   (c) $x = 4$

   (d) Fail to reject H_0.

   (e) There is not enough evidence at the 5% significance level to support the claim that the verbal SAT scores improved.

19. (a) H_0: The proportion of adults who prefer unplanned travel activities is equal to the proportion of adults who prefer planned travel activities. (claim)

   H_a: The proportion of adults who prefer unplanned travel activities is not equal to the proportion of adults who prefer planned travel activities.

   Critical value is 3.

   $\alpha = 0.05$

   $x = 5$

   Fail to reject H_0.

   (b) There is not enough evidence at the 5% significance level to reject the claim that the proportion of adults who prefer unplanned travel activities is equal to the proportions of adults who prefer planned travel activities.

21. (a) H_0: median $\leq \$585$ (claim); H_a: median $> \$585$

(b) Critical value is $z_0 = 2.33$.

(c) $x = 29$

$$z = \frac{(x - 0.5) - 0.5(n)}{\frac{\sqrt{n}}{2}} = \frac{(29 - 0.5) - 0.5(47)}{\frac{\sqrt{47}}{2}} = \frac{5}{3.428} \approx 1.46$$

(d) Fail to reject H_0.

(e) There is not enough evidence at the 1% significance level to reject the claim that the median weekly earnings of female workers is less than or equal to $585.

23. (a) H_0: median ≤ 25.5; H_a: median > 25.5 (claim)

(b) Critical value is $z_0 = 1.645$.

(c) $x = 38$

$$z = \frac{(x + 0.5) - 0.5(n)}{\frac{\sqrt{n}}{2}} = \frac{(38 + 0.5) - 0.5(60)}{\frac{\sqrt{60}}{2}} = \frac{8.5}{3.873} \approx 1.936$$

(d) Reject H_0.

(e) There is enough evidence at the 5% significance level to support the claim that the median age of first-time brides is greater than 25.5 years.

11.2 THE WILCOXON TESTS

11.2 Try It Yourself Solutions

1a. H_0: The water repellent is not effective.

H_a: The water repellent is effective. (claim)

b. $\alpha = 0.01$ **c.** $n = 11$

d. Critical value is 5.

e.

No repellent	Repellent applied	Difference	Absolute value	Rank	Signed rank
8	15	−7	7	11	−11
7	12	−5	5	9	−9
7	11	−4	4	7.5	−7.5
4	6	−2	2	3.5	−3.5
6	6	0	0	−	−
10	8	2	2	3.5	3.5
9	8	1	1	1.5	1.5
5	6	−1	1	1.5	−1.5
9	12	−3	3	5.5	−5.5
11	8	3	3	5.5	5.5
8	14	−6	6	10	−10
4	8	−4	4	7.5	−7.5

Sum of negative ranks $= -55.5$

Sum of positive ranks $= 10.5$

$w_s = 10.5$

f. Fail to reject H_0.

g. There is not enough evidence at the 1% significance level to support the claim.

2a. H_0: There is no difference in the claims paid by the companies.

   H_a: There is a difference in the claims paid by the companies. (claim)

b. $\alpha = 0.05$

c. The critical values are $z_0 = \pm 1.96$.

d. $n_1 = 12$ and $n_2 = 12$

e.

Ordered data	Sample	Rank		Ordered data	Sample	Rank
1.7	B	1		5.3	B	13
1.8	B	2		5.6	B	14
2.2	B	3		5.8	A	15
2.5	A	4		6.0	A	16
3.0	A	5.5		6.2	A	17
3.0	B	5.5		6.3	A	18
3.4	B	7		6.5	A	19
3.9	A	8		7.3	B	20
4.1	B	9		7.4	A	21
4.4	B	10		9.9	A	22
4.5	A	11		10.6	A	23
4.7	B	12		10.8	B	24

R = sum ranks of company B = 120.5

f. $\mu_R = \dfrac{n_1(n_1 + n_2 + 1)}{2} = \dfrac{12(12 + 12 + 1)}{2} = 150$

$\sigma_R = \sqrt{\dfrac{n_1 n_2(n_1 + n_2 + 1)}{12}} = \sqrt{\dfrac{(12)(12)(12 + 12 + 1)}{12}} \approx 17.321$

$z = \dfrac{R - \mu_R}{\sigma_R} = \dfrac{120.5 - 150}{17.321} \approx -1.703$

g. Fail to reject H_0.

h. There is not enough evidence to conclude that there is a difference in the claims paid by both companies.

11.2 EXERCISE SOLUTIONS

1. The Wilcoxon signed-rank test is used to determine whether two dependent samples were selected from populations having the same distribution. The Wilcoxon rank sum test is used to determine whether two independent samples were selected from populations having the same distribution.

3. (a) H_0: There is no reduction in diastolic blood pressure. (claim)

   H_a: There is a reduction in diastolic blood pressure.

   (b) Wilcoxon signed-rank test   (c) Critical value is 10.

   (d) $w_s = 17$ (e) Fail to reject H_0.

   (f) There is not enough evidence at the 1% significance level to reject the claim that there was no reduction in diastolic blood pressure.

5. (a) H_0: There is no difference in the earnings.

H_a: There is a difference in the earnings. (claim)

(b) Wilcoxon rank sum test

(c) The critical values are $z_0 = \pm 1.96$.

(d) $R = 58$

$$\mu_R = \frac{n_1(n_1 + n_2 + 1)}{2} = \frac{11(11 + 10 + 1)}{2} = 110$$

$$\sigma_R = \sqrt{\frac{n_1 n_2(n_1 + n_2 + 1)}{12}} = \sqrt{\frac{(11)(10)(11 + 10 + 1)}{12}} \approx 14.201$$

$$z = \frac{R - \mu_R}{\sigma_R} = \frac{58 - 110}{14.201} \approx -3.66$$

(e) Reject H_0.

(f) There is enough evidence at the 5% significance level to support the claim that there is a difference in the earnings.

7. (a) H_0: There is not a difference in salaries.

H_a: There is a difference in salaries. (claim)

(b) Wilcoxon rank sum test

(c) The critical values are $z_0 = \pm 1.96$.

(d) $R = 117$

$$\mu_R = \frac{n_1(n_1 + n_2 + 1)}{2} = \frac{12(12 + 12 + 1)}{2} = 150$$

$$\sigma_R = \sqrt{\frac{n_1 n_2(n_1 + n_2 + 1)}{12}} = \sqrt{\frac{(12)(12)(12 + 12 + 1)}{12}} \approx 17.321$$

$$z = \frac{R - \mu_R}{\sigma_R} = \frac{117 - 150}{17.321} \approx -1.91$$

(e) Fail to reject H_0.

(f) There is not enough evidence at the 5% significance level to support the claim that there is a difference in salaries.

9. H_0: The fuel additive does not improve gas mileage.

H_a: The fuel additive does improve gas mileage. (claim)

Critical value is $z_0 = 1.282$.

$$w_s = 43.5$$

$$z = \frac{w_s - \frac{n(n + 1)}{4}}{\sqrt{\frac{n(n + 1)(2n + 1)}{24}}} = \frac{43.5 - \frac{32(32 + 1)}{4}}{\sqrt{\frac{32(32 + 1)[(2)32 + 1]}{24}}} = \frac{-220.5}{\sqrt{2860}} \approx -4.123$$

Note: $n = 32$ because one of the differences is zero and should be discarded.

Reject H_0. There is enough evidence at the 10% level to conclude that the gas mileage is improved.

11.3 THE KRUSKAL-WALLIS TEST

11.3 Try It Yourself Solutions

1a. H_0: There is no difference in the salaries in the three states.

H_a: There is a difference in the salaries in the three states. (claim)

b. $\alpha = 0.10$

c. d.f. $= k - 1 = 2$

d. Critical value is $\chi_0^2 = 4.605$; Reject H_0 if $\chi^2 > 4.605$.

e.

Ordered data	State	Rank	Ordered data	State	Rank
86.85	MI	1	94.72	NC	16
87.25	NC	2	94.75	NC	17
87.70	CO	3	95.10	MI	18
89.65	MI	4	95.36	NC	19
89.75	CO	5	96.02	NC	20
89.92	NC	6	96.24	CO	21
91.17	CO	7	96.31	MI	22
91.55	CO	8	97.35	CO	23
92.85	CO	9	98.21	MI	24
93.12	CO	10	98.34	NC	25
93.76	MI	11	98.99	CO	26
93.92	MI	12	100.27	NC	27
94.42	MI	13	105.77	NC	28
94.45	MI	14	106.78	MI	29
94.55	CO	15	110.99	NC	30

$R_1 = 127 \quad R_2 = 148 \quad R_3 = 190$

f. $H = \dfrac{12}{N(N+1)}\left(\dfrac{R_1^2}{n_1} + \dfrac{R_2^2}{n_2} + \dfrac{R_3^2}{n_3}\right) - 3(N+1)$

$= \dfrac{12}{30(30+1)}\left(\dfrac{(127)^2}{10} + \dfrac{(148)^2}{10} + \dfrac{(190)^2}{10}\right) - 3(30+1) = 2.655$

g. Fail to reject H_0.

h. There is not enough evidence to support the claim.

11.3 EXERCISE SOLUTIONS

1. Each sample must be randomly selected and the size of each sample must be at least 5.

3. (a) H_0: There is no difference in the premiums.

H_a: There is a difference in the premiums. (claim)

(b) Critical value is 5.991.

(c) $H \approx 13.091$

(d) Reject H_0.

(e) There is enough evidence at the 5% significance level to support the claim that the distributions of the annual premiums in Arizona, Florida, and Louisiana are different.

5. (a) H_0: There is no difference in the salaries.

H_a: There is a difference in the salaries. (claim)

(b) Critical value is 6.251.

(c) $H \approx 1.024$

(d) Fail to reject H_0.

(e) There is not enough evidence at the 10% significance level to support the claim that the distributions of the annual salaries in the four states are different.

7. (a) H_0: There is no difference in the number of days spent in the hospital.

H_a: There is a difference in the number of days spent in the hospital. (claim)

The critical value is 11.345.

$H = 1.51$;

Fail to reject H_0.

(b)

Variation	Sum of squares	Degrees of freedom	Mean Squares	F
Between	9.17	3	3.06	0.52
Within	194.72	33	5.90	

For $\alpha = 0.01$, the critical value is about 4.45. Because $F = 0.52$ is less than the critical value, the decision is to fail to reject H_0. There is not enough evidence to support the claim.

(c) Both tests come to the same decision, which is that there is not enough evidence to support the claim that there is a difference in the number of days spent in the hospital.

11.4 RANK CORRELATION

11.4 Try It Yourself Solutions

1a. $H_0: \rho_s = 0$; $H_a: \rho_s \neq 0$

b. $\alpha = 0.05$

c. Critical value is 0.700.

d.

Oat	Rank	Wheat	Rank	d	d^2
1.10	1.5	2.65	3	−1.5	2.25
1.12	3	2.48	1	2	4
1.10	1.5	2.62	2	−0.5	0.25
1.59	6	2.78	4	2	4
1.81	8	3.56	8	0	0
1.48	4.5	3.40	5.5	−1	1
1.48	4.5	3.40	5.5	−1	1
1.63	7	4.25	7	0	0
1.85	9	4.25	9	0	0
					$\Sigma = 12.5$

$\Sigma d^2 = 12.5$

e. $r_s = 1 - \dfrac{6\Sigma d^2}{n(n^2 - 1)} = 0.896$

f. Reject H_0.

g. There is enough evidence to conclude that a significant correction exists.

11.4 EXERCISE SOLUTIONS

1. The Spearman rank correlation coefficient can (1) be used to describe the relationship between linear and nonlinear data, (2) be used for data at the ordinal level, and (3) is easier to calculate by hand than the Pearson coefficient.

3. (a) $H_0: \rho_s = 0; H_a: \rho_s \neq 0$ (claim)

(b) Critical value is 0.929.

(c) $\Sigma d^2 = 8$

$$r_s = 1 - \dfrac{6\Sigma d^2}{n(n^2 - 1)} \approx 0.857$$

(d) Fail to reject H_0.

(e) There is not enough evidence at the 1% significance level to support the claim that there is a correlation between debt and income in the farming business.

5. (a) $H_0: \rho_s = 0; H_a: \rho_s \neq 0$ (claim)

(b) The critical value is 0.881.

(c) $\Sigma d^2 = 66.5$

$$r_s = 1 - \dfrac{6\Sigma d^2}{n(n^2 - 1)} \approx 0.208$$

(d) Fail to reject H_0.

(e) There is not enough evidence at the 1% significance level to support the claim that there is a correlation between the overall score and price.

7. $H_0: \rho_s = 0; H_a: \rho_s \neq 0$ (claim)

Critical value is 0.700.

$\Sigma d^2 = 121.5$

$$r_s = 1 - \dfrac{6\Sigma d^2}{n(n^2 - 1)} \approx -0.013$$

Fail to reject H_0. There is not enough evidence at the 5% significance level to conclude that there is a correlation between science achievement scores and GNP.

9. $H_0: \rho_s = 0; H_a: \rho_s \neq 0$ (claim)

Critical value is 0.700.

$\Sigma d^2 = 9.5$

$$r_s = 1 - \dfrac{6\Sigma d^2}{n(n^2 - 1)} \approx 0.921$$

Reject H_0. There is enough evidence at the 5% significance level to conclude that there is a correlation between science and mathematics achievement scores.

11. $H_0: \rho_s = 0; H_a: \rho_s \neq 0$ (claim)

The critical value is $= \dfrac{\pm z}{\sqrt{n-1}} = \dfrac{\pm 1.96}{\sqrt{33-1}} \approx \pm 0.346$.

$\Sigma d^2 = 6673$

$r_s = 1 - \dfrac{6\Sigma d^2}{n(n^2-1)} \approx -0.115$

Fail to reject H_0. There is not enough evidence to support the claim.

11.5 THE RUNS TEST

11.5 Try It Yourself Solutions

1a. *P P P F P F P P P P F F P F P P F F F P P P F P P P*

b. 13 groups $\Rightarrow$ 13 runs

c. 3, 1, 1, 1, 4, 2, 1, 1, 2, 3, 3, 1, 3

2a. H_0: The sequence of genders is random.

H_a: The sequence of genders is not random. (claim)

b. $\alpha = 0.05$

c. *F F F M M F F M F M M F F F*

$n_1 = $ number of *F*s $= 9$

$n_2 = $ number of *M*s $= 5$

$G = $ number of runs $= 7$

d. $cv = 3$ and 12

e. $G = 7$

f. Fail to reject H_0.

g. At the 5% significance level, there is not enough evidence to support the claim that the sequence of genders is not random.

3a. H_0: The sequence of weather conditions is random.

H_a: The sequence of weather conditions is not random. (claim)

b. $\alpha = 0.05$

c. $n_1 = $ number of *N*s $= 21$

$n_2 = $ number of *S*s $= 10$

$G = $ number of runs $= 17$

d. $cv = \pm 1.96$

e. $\mu_G = \dfrac{2n_1 n_2}{n_1 + n_2} + 1 = \dfrac{2(21)(10)}{21 + 10} + 1 \approx 14.5$

$\sigma_G = \sqrt{\dfrac{2n_1 n_2(2n_1 n_2 - n_1 - n_2)}{(n_1 + n_2)^2(n_1 + n_2 - 1)}} = \sqrt{\dfrac{2(21)(10)(2(21)(10) - 21 - 10)}{(21 + 10)^2(21 + 10 - 1)}} \approx 2.4$

$z = \dfrac{G - \mu_G}{\sigma_G} = \dfrac{17 - 14.5}{2.4} = 1.04$

f. Fail to reject H_0.

g. At the 5% significance level, there is not enough evidence to support the claim that the sequence of weather conditions each day is not random.

11.5 EXERCISE SOLUTIONS

1. Number of runs = 8
Run lengths = 1, 1, 1, 1, 3, 3, 1, 1

3. Number of runs = 9
Run lengths = 1, 1, 1, 1, 1, 6, 3, 2, 4

5. n_1 = number of Ts = 6
n_2 = number of Fs = 6

7. n_1 = number of Ms = 10
n_2 = number of Fs = 10

9. n_1 = number of Ts = 6
n_1 = number of Fs = 6

cv = 3 and 11

11. n_1 = number of Ns = 11
n_1 = number of Ss = 7

cv = 5 and 14

13. (a) H_0: The coin tosses were random.
H_a: The coin tosses were not random. (claim)

(b) n_1 = number of Hs = 7
n_2 = number of Ts = 9

cv = 4 and 14

(c) G = 9 runs

(d) Fail to reject H_0.

(e) At the 5% significance level, there is not enough evidence to support the claim that the coin tosses were not random.

15. (a) H_0: The sequence of digits was randomly generated.
H_a: The sequence of digits was not randomly generated. (claim)

(b) n_1 = number of Os = 16
n_2 = number of Es = 16

cv = 11 and 23

(c) G = 9 runs

(d) Reject H_0.

(e) At the 5% significance level, there is enough evidence to support the claim that the sequence of digits was not randomly generated.

17. (a) H_0: The sequence is random.

H_a: The sequence is not random. (claim)

(b) n_1 = number of Ns = 40

n_2 = number of Ps = 9

$cv = \pm 1.96$

(c) G = 14 runs

$$\mu_G = \frac{2n_1n_2}{n_1 + n_2} + 1 = \frac{2(40)(9)}{40 + 9} + 1 \approx 15.7$$

$$\sigma_G = \sqrt{\frac{2n_1n_2(2n_1n_2 - n_1 - n_2)}{(n_1 + n_2)^2(n_1 + n_2 - 1)}} = \sqrt{\frac{2(40)(9)(2(40)(9) - 40 - 9)}{(40 + 9)^2(40 + 9 - 1)}} = 2.05$$

$$z = \frac{G - \mu_G}{\sigma_G} = \frac{14 - 15.7}{2.05} = -0.83$$

(d) Fail to reject H_0.

(e) At the 5% significance level, there is not enough evidence to support the claim that the sequence is not random.

19. H_0: Daily high temperatures occur randomly.

H_a: Daily high temperatures do not occur randomly. (claim)

median = 87

n_1 = number above median = 14

n_2 = number below median = 13

cv = 9 and 20

G = 11 runs

Fail to reject H_0.

At the 5% significance level, there is not enough evidence to support the claim that the daily high temperatures do not occur randomly.

21. Answers will vary.

CHAPTER 11 REVIEW EXERCISE SOLUTIONS

1. (a) H_0: median = $24,300 (claim); H_a: median $\neq$ $24,300

(b) Critical value is 2.

(c) $x = 7$

(d) Fail to reject H_0.

(e) There is not enough evidence at the 1% significance level to reject the claim that the median value of stock among families that own stock is $24,300.

3. (a) H_0: median ≤ 6 (claim); H_a: median > 6

(b) Critical value is $z_0 \approx -1.28$.

(c) $x = 44$

$$z = \frac{(x - 0.5) - 0.5(n)}{\frac{\sqrt{n}}{2}} = \frac{(44 - 0.5) - 0.5(70)}{\frac{\sqrt{70}}{2}} = \frac{8.5}{4.1833} \approx 2.03$$

(d) Reject H_0.

(e) There is enough evidence at the 10% significance level to reject the claim that the median turnover time is no more than 6 hours.

5. (a) H_0: There is no reduction in diastolic blood pressure. (claim)

H_a: There is a reduction in diastolic blood pressure.

(b) Critical value is 2.

(c) $x = 3$

(d) Fail to reject H_0.

(e) There is not enough evidence at the 5% significance level to reject the claim that there was no reduction diastolic blood pressure.

7. (a) Independent; Wilcoxon Rank Sum Test

(b) H_0: There is no difference in the amount of time that it takes to earn a doctorate.

H_a: There is a difference in the amount of time that it takes to earn a doctorate. (claim)

(c) Critical values are $z_0 = \pm 2.575$.

(d) $R = 173.5$

$$\mu_R = \frac{n_1(n_1 + n_2 + 1)}{2} = \frac{12(12 + 12 + 1)}{2} = 150$$

$$\sigma_R = \sqrt{\frac{n_1 n_2(n_1 + n_2 + 1)}{12}} \sqrt{\frac{(12)(12)(12 + 12 + 1)}{12}} \approx 17.321$$

$$z = \frac{R - \mu_R}{\sigma_R} = \frac{173.5 - 150}{17.321} \approx 1.357$$

(e) Fail to reject H_0.

(f) There is not enough evidence at the 1% significance level to support the claim that there is a difference in the amount of time that it takes to earn a doctorate.

9. (a) H_0: There is no difference in salaries between the fields of study.

H_a: There is a difference in salaries between the fields of study. (claim)

(b) Critical value is 5.991.

(c) $H \approx 21.695$

(d) Reject H_0.

(e) There is enough evidence at the 5% significance level to conclude that there is a difference in salaries between the fields of study.

11. (a) $H_0: \rho_s = 0; H_a: \rho_s \neq 0$ (claim)

(b) Critical value is 0.881.

(c) $\Sigma d^2 = 57$

$$r_s = 1 - \frac{6\Sigma d^2}{n(n^2 - 1)} \approx 0.321$$

(d) Fail to reject H_0.

(e) There is not enough evidence at the 1% significance level to support the claim that there is a correlation between overall score and price.

13. (a) H_0: The traffic stops were random by gender.

H_a: The traffic stops were not random by gender. (claim)

(b) $n_1 = $ number of Fs $= 12$

$n_2 = $ number of Ms $= 13$

$cv = 8$ and 19

(c) $G = 14$ runs

(d) Fail to reject H_0.

(e) There is not enough evidence at the 5% significance level to support the claim that the traffic stops were not random.

CHAPTER 11 QUIZ SOLUTIONS

1. (a) H_0: There is no difference in the salaries between genders.

H_a: There is a difference in the salaries between genders. (claim)

(b) Wilcoxon Rank Sum Test

(c) Critical values are $z_0 = \pm 1.645$.

(d) $R = 67.5$

$$\mu_R = \frac{n_1(n_1 + n_2 + 1)}{2} = \frac{10(10 + 10 + 1)}{2} = 105$$

$$\sigma_R = \sqrt{\frac{n_1 n_2(n_1 + n_2 + 1)}{12}} = \sqrt{\frac{(10)(10)(10 + 10 + 1)}{12}} \approx 13.229$$

$$z = \frac{R - \mu_R}{\sigma_R} = \frac{67.5 - 105}{13.229} \approx -2.835$$

(e) Reject H_0.

(f) There is enough evidence at the 10% significance level to support the claim that there is a difference in the salaries between genders.

2. (a) H_0: median $= 50$ (claim); H_a: median $\neq 50$

(b) Sign Test   (c) Critical value is 5.

(d) $x = 9$ (e) Fail to reject H_0.

(f) There is not enough evidence at the 5% significance level to reject the claim that the median number of annual volunteer hours is 50.

3. (a) H_0: There is no difference in rent between regions.

   H_a: There is a difference in rent between regions. (claim)

(b) Kruskal-Wallis Test   (c) Critical value is 7.815.

(d) $H \approx 11.826$ (e) Reject H_0.

(f) There is enough evidence at the 5% significance level to conclude that there is a difference in rent between regions.

4. (a) H_0: The days with rain are random.

   H_a: The days with rain are not random. (claim)

(b) The Runs test

(c) n_1 = number of Ns = 15
   n_2 = number of Rs = 15

   cv = 10 and 22

(d) G = 16 runs

(e) Fail to reject H_0.

(f) There is not enough evidence at the 5% significance level to conclude that days with rain are not random.

CUMULATIVE REVIEW FOR CHAPTERS 9–11

1. (a)

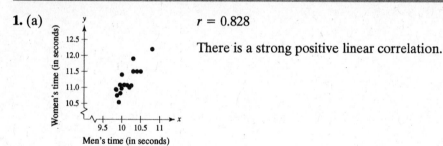

$r = 0.828$

There is a strong positive linear correlation.

(b) H_0: $\rho = 0$

   H_a: $\rho \neq 0$ (claim)

   $t = 5.911$

   $p = 0.0000591$

   Reject H_0. There is enough evidence at the 5% significance level to conclude that there is a significant linear correlation.

(c) $\hat{y} = 1.423x - 3.204$

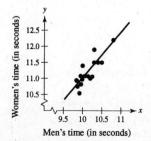

(d) $\hat{y} = 1.423(9.90) - 3.204 = 10.884$ seconds

2. H_0: Median (Union) = Median (Non union)

H_1: Median (Union) $\neq$ Median (Non union) (claim)

$cv = \pm 1.96$

$R = 145.0$

$$\mu_R = \frac{n_1(n_1 + n_2 + 1)}{2} = \frac{10(10 + 10 + 1)}{2} = 105$$

$$\sigma_R = \sqrt{\frac{n_1 n_2(n_1 + n_2 + 1)}{12}} = \sqrt{\frac{10(10)(10 + 10 + 1)}{12}} = 13,229$$

$$z = \frac{R - \mu_R}{\sigma_R} = \frac{145.0 - 105}{13,229} = 4.375$$

Reject H_0. There is enough evidence at the 5% significance level to support the claim.

3. H_0: median = 48 (claim)

H_1: median $\neq$ 48

$cv = 3$

$x = 7$

Fail to reject H_0. There is not enough evidence at the 5% significance level to reject the claim.

4. H_0: $\mu_1 = \mu_2 = \mu_3 = \mu_4$ (claim)

H_1: At least one μ if different.

$cv = 2.29$

$F = 2.476$

Reject H_0. There is enough evidence at the 10% significance level to reject the claim.

5. H_0: $\sigma_1^2 = \sigma_2^2$ (claim)

H_1: $\sigma_1^2 \neq \sigma_2^2$

$cv = 2.01$

$$F = \frac{s_1^2}{s_2^2} = \frac{34.6}{33.2} = 1.042$$

Fail to reject H_0. There is not enough evidence at the 10% significance level to reject the claim.

6. H_0: The medians are all equal.

H_1: The medians are not all equal. (claim)

$cv = 11.345$

$H = 14.78$

Reject H_0. There is enough evidence at the 1% significance level to support the claim.

7.

Physicians	Distribution
1	0.36
2–4	0.32
5–9	0.20
10+	0.12

H_0: The distribution is as claimed. (claim)

H_1: The distribution is not as claimed.

Physician	Distribution	Observed	Expected	$\frac{(O - E)^2}{E}$
1	0.36	116	104.04	1.375
2–4	0.32	84	92.48	0.778
5–9	0.20	66	57.80	1.163
10+	0.12	23	34.68	3.934
		289		$x^2 = 7.250$

$cv = 7.815$

$x^2 = 7.250$

Fail to reject H_0. There is not enough evidence at the 5% significance level to reject the claim.

8. (a) $r^2 = 0.733$

Metacarpal bone length explains 73.3% of the variability in height. About 26.7% of the variation is unexplained.

(b) $s_e = 4.255$

(c) $\hat{y} = 94.428 + 1.700(50) = 179.428$

$$E = t_c S_e \sqrt{1 + \frac{1}{n} + \frac{n(x_0 - \bar{x})^2}{n\Sigma x^2 - (\Sigma x)^2}} = 2.365(4.255)\sqrt{1 + \frac{1}{9} + \frac{9(50 - 45,444)^2}{9(18,707) - (409)^2}}$$
$$= 2.365(4.255)\sqrt{1.284} = 11.402$$

$\hat{y} \pm E \Longrightarrow (168.026, 190.83)$

You can be 95% confident that the height will be between 168.026 centimeters and 190.83 centimeters when the metacarpal bone length is 50 centimeters.

9.

Score	Rank	Price	Rank	d	d^2
85	8	81	7	1.0	1.00
83	6.5	78	6	0.5	0.25
83	6.5	56	1	5.5	30.25
81	5	77	5	0	0.00
77	4	62	3.5	0.5	0.25
72	3	85	8	−5.0	25.00
70	2	62	3.5	−1.5	2.25
66	1	61	2	−1.0	1.00
					$\Sigma d^2 = 60$

H_0: $\rho_s = 0$

H_1: $\rho_s \neq 0$ (claim)

$cv = 0.643$

$$r_s = 1 - \frac{6\Sigma d^2}{n(n^2 - 1)} = 1 - \frac{60(60)}{8(8^2 - 1)} = 0.286$$

Fail to reject H_0. There is not enough evidence at the 10% significance level to support the claim.

10. (a) $\hat{y} = 91.113 - 0.014(4325) + 0.018(1900) = 64.763$ bushels

(b) $\hat{y} = 91.113 - 0.014(4900) + 0.018(2163) = 61.447$ bushels

Alternative Presentation of the Standard Normal Distribution

Try It Yourself Solutions

1 (1) 0.4857

(2) $z = \pm 2.17$

2a.

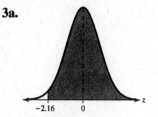

b. 0.4834

c. Area = 0.5 + 0.4834 = 0.9834

3a.

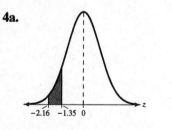

b. 0.4846

c. Area = 0.5 + 0.4846 = 0.9846

4a.

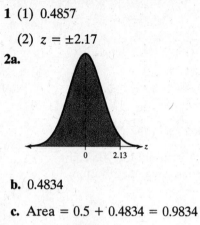

b. $z = -2.16$: Area = 0.4846

$z = -1.35$: Area = 0.4115

c. Area = 0.4846 − 0.4115 = 0.0731

Try It Yourself Solutions

1a.

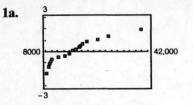

The points do not appear to be approximately linear.

b. 39,860 is a possible outlier because it is far removed from the other entries in the data set.

c. Because the points do not appear to be approximately linear and there is an outlier, you can conclude that the sample data do not come from a population that has a normal distribution.